EXERCISE 8 • CLIMATE CLASSIFICATION AND REGIONAL CLIMATES

PURPOSE

The purpose of this exercise is to learn how climate data is organized to create a world classification of climate, and to gain an understanding of the influence of continentality, topography, and ocean currents on the spatial patterns of climate.

LEARNING OBJECTIVES

By the end of this exercise you should be able to

- use the Köppen system to classify a climate;
- describe differences in seasonal temperature and precipitation patterns for different climate types and explain factors responsible for these differences;
- explain how ocean currents affect air stability and precipitation; and
- explain how mountains affect spatial patterns of precipitation and climate type.

INTRODUCTION

Examining a map of the world distribution of climates reveals a mosaic of climates spread across the earth. The geography of earth's climate is the evolutionary result of billions of years of earth history and is created by the changing composition of the atmosphere, the shifting of continents, biologic regulators, and astronomical events such as the tilt of the earth's axis, and meteors. If one looks closely at this apparent hodge-podge of climates, subtle regularities reveal themselves and point to the causal factors that created them. The first step in determining the cause of climate patterns requires us to organize a vast amount of information in a sensible and easily communicated way.

Global Climate Controls

The present-day climate of a region is the result of the complex interaction of location, wind and pressure systems, water bodies, and topography. The latitude of a place determines the amount of solar radiation received, because latitude affects sun angle and day length. Latitude also determines which global wind or pressure systems are dominant. The tropical rainy climates are dominated by the rain-producing **intertropical convergence zone.** The tropical dry climates are dominated by the rain-suppressing **subtropical highs.** The seasonal precipitation typical of some subtropical climates is due to the shifting presence of the subtropical high- and **subpolar low**-pressure cells. For example, the dry summer subtropical climate, sometimes called the Mediterranean climate, experiences its dry period when the subtropical high dominates. The rainy period occurs when the subpolar low dominates. The variable climate conditions of the midlatitudes are in part a reflection of the migratory nature of the **polar front jet stream** and the subpolar low.

Ocean currents and mountain systems are elements that provide important regional controls of climate. Surface winds created by the subtropical highs are responsible for most of the major ocean currents, which have a moderating effect on the temperature of a coastal region. That is, ocean currents tend to decrease the range of temperatures experienced by a place. Recall that large bodies of water, like oceans, are **heat reservoirs,** absorbing heat slowly and retaining it for long periods of time. Thus, water temperatures change quite slowly, as does the air temperature above water. Oceanic air will not heat up or cool down as much as will air sitting above a land surface.

In addition to affecting temperature, ocean currents can have an extreme effect on the amount of moisture a coastal region receives. Cold ocean currents act to stabilize the air at the surface, thus preventing uplift and precipitation. Warm ocean currents can help promote instability in the air, enhancing uplift and precipitation. **Upwelling** of cold water (from the lower depths of the ocean) enhances the effect of a cold water current. Upwelling is especially strong along coasts with a steep continental shelf. This helps explain why water off the west coast of South America is so cool even near the equator.

Since ocean currents essentially parallel the direction of the global wind system, if we know the direction of the winds diverging around the subtropical highs, we can determine where the major oceans currents flow. Likewise, we can determine which ocean currents are cold and which are warm. Currents flowing away from the Poles are cold, whereas currents flowing away from the equator are warm.

The orientation of a mountain range with respect to the prevailing winds of a region can play a significant role in influencing the weather and climate of a region. Midlatitude weather systems tend to move in a west-to-east direction in the Northern Hemisphere. Most major mountain systems in North and South America show a north-south orientation while many European and Asian mountain systems are oriented in an east-west direction. As a result, maritime air masses penetrate farther inland in Europe than they do in North America. This creates "oceanic" climate conditions (high average humidity, high rainfall, and small temperature ranges) farther into the interior of Europe than in North America. The north-south aligned mountains in North America cause uplift and precipitation on their windward western slopes and dry conditions on the leeward eastern slopes.

Climate Classification

The geographic distribution of atmospheric phenomena is complex to say the least. The spatial and temporal variation of climate elements (e.g., air temperature, precipitation, air masses, areas of cyclogenesis, topography) results in place-to-place variations of environmental conditions. Such complex variation makes the study of climate quite difficult. This requires the adoption of a framework for arranging the vast information concerning climate, that is, a classification of climate.

Classification is a fundamental tool of science. The objectives of classification are three-fold: (1) to organize large quantities of information, (2) to speed retrieval of information, and (3) to facilitate communication of information. Climate classification is concerned with the organization of climatic data in such a way that both descriptive and analytical generalizations can be made. In addition, it attempts to store information in an orderly fashion for easy reference and communication, often in the form of maps, graphs, or tables. The usefulness of any particular classification system is largely determined by its intended use.

There are three basic types of classification systems—empirical, genetic, and applied. An **empirical classification** system classifies climates solely on the basis of observable features (e.g., temperature, precipitation). A **genetic classification** system classifies climates according to the cause of observable features (e.g., frequency of air mass invasions, influence of orographic barriers, influence of particular wind and pressure belts). **Applied classification** systems assist in the solution of specialized problems. Explanations of the distribution of natural vegetation or assessment of the distribution of soil moisture availability are two examples where applied classifications have been formulated. Applied classifications have used precipitation, potential evapotranspiration, and indices of moisture adequacy or thermal efficiency as classification criteria.

Köppen Climate Classification

The climate classification used in this exercise was developed by Wladimir Köppen and is based upon annual and monthly means of temperature and precipitation. The Köppen classification identifies five main groups of climates (designated by capital letters), all but one, the dry group, being thermally defined. They are as follows:

A-type climates: Tropical rainy climates; hot all seasons

B-type climates: Dry climates

C-type climates: Warm temperate rainy climates; mild winters

D-type climates: Boreal climates; severe winters

E-type climates: Polar climates

A sixth group for highland climate (H) is also recognized but has no temperature or precipitation criteria. Highland climate is that in which climate conditions change significantly over a short horizontal distance due to altitudinal variations.

In order to represent the main climate types, additional letters are used. The second letter pertains to precipitation characteristics and the third to temperature. Table 8.1 describes the main climate types and Table 8.2 explains the symbols and criteria utilized for subdividing the main climate types.

Procedure for Classifying Climate

The procedure for determining the climate of a place should be worked through step-by-step.

1. Determine if the station is an E-type climate (refer to Table 8.2). If it is, go to step 2; if not, proceed to step 3.

2. Determine if the station is an ET or EF climate (see Table 8.2).

3. Determine if the station is a dry climate. If the station has a dry climate, identify the appropriate subcategories. If it is not an arid climate, continue on to step 4. Use Figure 8.1 if 70% of precipitation falls during the summer. Use Figure 8.2 if 70% of precipitation falls during the winter. Note that the summer season is April through September for the Northern Hemisphere, while the winter season is October through March. This is reversed for the Southern Hemisphere. Use Figure 8.3 if neither half of year has more than 70% of annual precipitation.

4. Determine if the station is an A-type climate (refer to Table 8.2). If so, go to step 5; if not, skip to step 6.

5. Determine subcategory of A-type climate, using Figure 8.4.

6. Determine if the station is a C- or D-type climate (refer to Table 8.2). Continue to step 7.

7. Identify the appropriate subcategories of the C- or D-type climate (Table 8.2).

TABLE 8.1	MAIN CLIMATE TYPES
Af	Tropical rainforest. Hot; rainy through all seasons.
Am	Tropical monsoon. Hot; pronounced wet season.
Aw	Tropical savanna. Hot; pronounced dry season (usually the low sun period).
BSh	Tropical steppe. Semiarid; hot.
BSk	Midlatitude steppe. Semiarid; cool or cold.
BWh	Tropical desert. Arid; hot.
BWk	Midlatitude desert. Arid; cool or cold.
Cfa	Humid subtropical. Mild winter; moist through all seasons; long, hot summer.
Cfb	Marine. Mild winter; moist through all seasons; warm summer.
Cfc	Marine. Mild winter; moist through all seasons; short, cool summer.
Csa	Interior Mediterranean. Mild winter; dry summer; hot summer.
Csb	Coastal Mediterranean. Mild winter; dry summer; cool summer.
Cwa	Subtropical monsoon. Mild winter; dry winter; hot summer.
Cwb	Tropical upland. Mild winter; dry winter; short, warm summer.
Dfa	Humid continental. Severe winter; moist through all seasons; long, hot summer.
Dfb	Humid continental. Severe winter; moist through all seasons; short, warm summer.
Dfc	Subarctic. Severe winter; moist through all seasons; short, cool summer.
Dfd	Subarctic. Extremely cold winter; moist through all seasons; short summer.
Dwa	Humid continental. Severe winter; dry winter; long, hot summer.
Dwb	Humid continental. Severe winter; dry winter; warm summer.
Dwc	Subarctic. Severe winter; dry winter; short, cool summer.
Dwd	Subarctic. Extremely cold winter; dry winter, short, cool summer.
ET	Tundra. Very short summer.
EF	Perpetual ice and snow.
H	Undifferentiated highland climates.

TABLE 8.2 SIMPLIFIED KÖPPEN CLASSIFICATION OF CLIMATE					
First Letter		**Second Letter**		**Third Letter**	
E	Warmest month <10°C	T	Warmest month between 10° – 0°C	Not applicable	
		F	Warmest month <0°C		
B	Arid and semiarid climate	S	Semiarid climate; see Figures 8.1, 8.2, or 8.3.	h	Mean annual temperature is >18°C
		W	Arid climate; see Figures 8.1, 8.2, or 8.3.	k	Mean annual temperature is <18°C
A	Coolest month >18°C	f	Driest month precipitation ≥60mm	Not applicable	
		m	Seasonally, excessively moist; see Figure 8.4.		
		w	Driest month in winter half of year; precipitation in driest month ≤60mm.		
C	Coldest month 18° – 0°C; at least one month >10°C	s	Driest month in summer half of year with ≤40mm of precipitation and <$1/3$ of the wettest winter month.	a	Warmest month ≥22°C
		w	Driest month in winter half of year with <$1/10$ precipitation of wettest summer month.	b	Warmest month <22°C with at least 4 months >10°C
		f	Does not meet criteria for s or w above (moist all year)	c	Warmest month <22°C with 1 – 3 months >10°C
D	Coldest month <0°C; at least one month >10°C	s	Same as above	a	Same as above
		w	Same as above	b	Same as above
		f	Same as above	c	Same as above
				d	Coldest month < – 38°C
H	Highland	Not applicable		Not applicable	

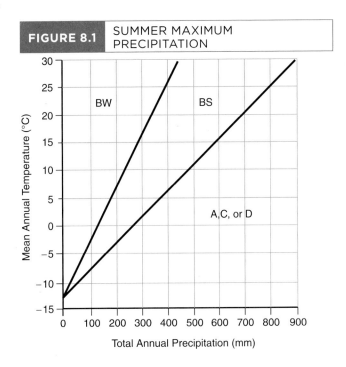

FIGURE 8.1 SUMMER MAXIMUM PRECIPITATION

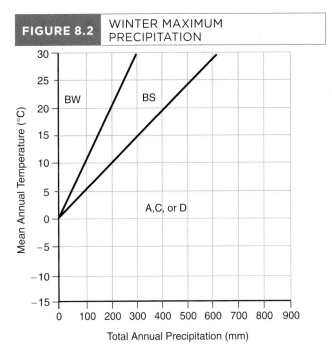

FIGURE 8.2 WINTER MAXIMUM PRECIPITATION

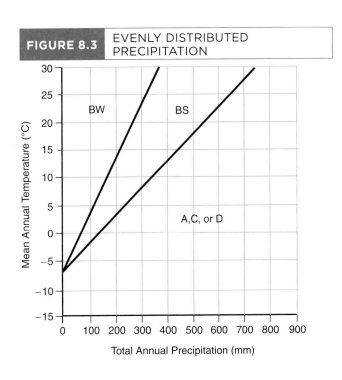

FIGURE 8.3 EVENLY DISTRIBUTED PRECIPITATION

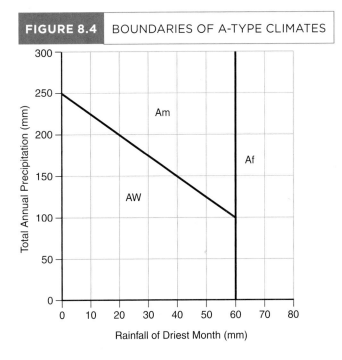

FIGURE 8.4 BOUNDARIES OF A-TYPE CLIMATES

IMPORTANT TERMS, PHRASES, AND CONCEPTS

- intertropical convergence zone
- subtropical high
- subpolar low
- polar front jet stream
- heat reservoirs
- upwelling
- empirical classification
- genetic classification
- applied classification

75

Name: _____   Section: _____

PART 1 † CLIMATE CLASSIFICATION AND DISTRIBUTION

1. Plot the stations listed in Table 8.3 on the world map provided in Figure 8.5.

TABLE 8.3 CLIMATE DATA FOR SELECTED CITIES

	Jan	Feb	Mar	Apr	May	Jun	Jul	Aug	Sep	Oct	Nov	Dec
Iquitos, Peru (3°S, 73°W)												
T (°C)	25.6	25.6	24.4	25	24.4	23.3	23.3	24.4	24.4	25	25.6	25.6
P (mm)	259	249	310	165	254	188	168	117	221	183	213	292
Chennai, India (13°N, 80°E)												
T (°C)	24.5	25.8	27.9	30.5	32.7	32.5	30.7	30.1	29.7	28.1	26.0	24.6
P (mm)	24	7	15	25	52	53	83	124	118	267	308	157
Rangoon, Myanmar (17°N, 96°E)												
T (°C)	24.3	25.2	27.2	29.8	29.5	27.8	27.6	27.1	27.6	28.3	27.7	25.0
P (mm)	8	5	6	17	260	524	492	574	398	208	34	3
Faya, Chad (18°N, 21°E)												
T (°C)	20.3	22.5	26.2	30.3	33.4	34.1	33.3	32.7	32.6	29.8	24.5	21.2
P (mm)	0	0	0	0	.5	1.1	4.4	10.9	.9	0	0	0
Colorado Springs, Colorado (39°N, 105°W)												
T (°C)	−1.7	0.0	2.9	8.0	13.0	18.3	21.5	20.1	15.7	9.9	3.2	−1.2
P (mm)	7.5	9.0	21.9	31.5	56.8	53.5	74.0	70.2	32.6	20.8	12.8	9.5
Sacramento, California (39°N, 122°W)												
T (°C)	8	10	12	16	19	22	25	24	23	18	12	9
P (mm)	81	76	60	36	15	3	—	1	5	20	37	82

(continued)

77

TABLE 8.3	CLIMATE DATA FOR SELECTED CITIES (*CONTINUED*)											
	Jan	Feb	Mar	Apr	May	Jun	Jul	Aug	Sep	Oct	Nov	Dec
Montgomery, Alabama (32°N, 86°W)												
T (°C)	7.9	10	13.9	18.2	22.3	26.1	27.6	27.3	24.7	18.6	13.2	9.3
P (mm)	109.6	129.4	152.6	114.8	100.9	94.3	126.7	85.2	113.9	61.3	92.4	128.9
Dubuque, Iowa (42°N, 90°W)												
T (°C)	−8.9	−6.0	1.0	8.6	14.7	19.8	22.3	21	16.4	10.2	2.3	−5.7
P (mm)	33.6	31.7	70.6	98.9	107.5	103.9	106.6	113.7	104.5	66.4	65.3	46
Barrow, Alaska (71°N, 156°W)												
T (°C)	−25.6	−27.6	−26.1	−19	−7	1.1	4	3.2	−0.8	−10.2	−18.7	−24
P (mm)	4.2	3.8	3.3	3.7	3.3	7.8	21.8	22.8	14.7	12.8	6.4	4.5
Albuquerque, New Mexico (35°N, 106°W)												
T (°C)	1.2	4.4	8.2	12.8	17.8	23.4	25.8	24.3	20.3	13.8	6.8	1.8
P (mm)	10.3	10.5	12.4	11.1	13.8	13.3	33.4	39.3	23.3	21.5	11.7	13.3
Greenwich, England (51°N, 0°)												
T (°C)	3.9	4.2	5.7	8.5	11.9	15.2	17	16.6	14.2	10.3	6.6	4.8
P (mm)	48.9	38.8	39.3	41.4	47	48.3	59	59.6	52.4	65.2	59.3	51.2
Dawson, Canada (64°N, 139°W)												
T (°C)	−28.6	−23.8	−14.5	−1.6	7.8	13.7	15.3	12.5	6	−3.5	−17.2	−24.9
P (mm)	19	16	12.1	11.1	23.7	34.2	45.1	43.2	33.3	29.2	25.5	24.2

2. Figure 8.6 contains climographs for all but four of the cities listed in Table 8.3. On the blank climographs provided in Figure 8.6, construct climographs for these four cities—Iquitos, Colorado Springs, Dubuque, and Dawson. Plot temperature as a line graph and precipitation as a bar graph.

3. Compute and enter the following statistics at the bottom of each climograph.

    a. Temperature range

    b. Average annual temperature

    c. Total annual precipitation

    d. Total summer precipitation

    e. Total winter precipitation

FIGURE 8.5 WORLD MAP

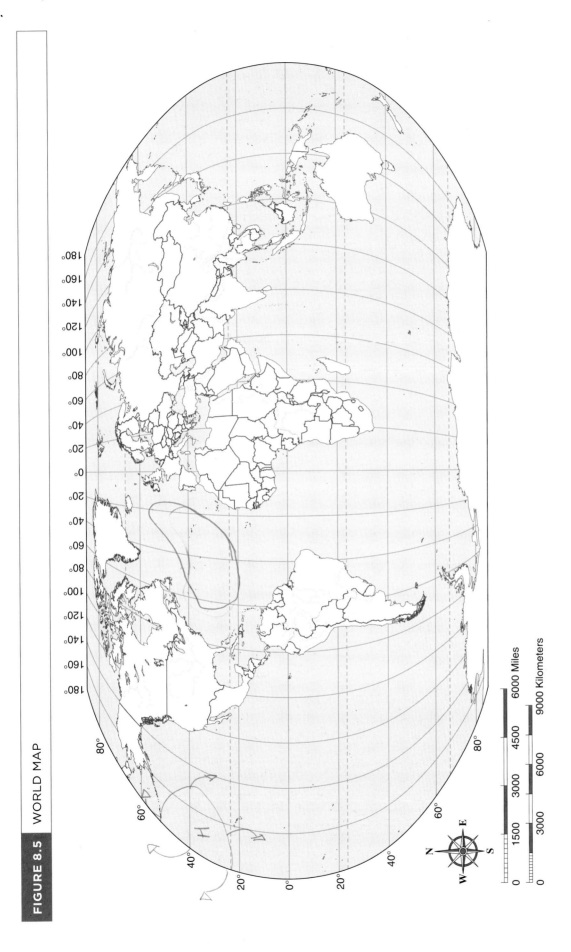

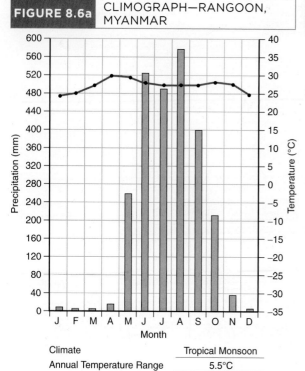

FIGURE 8.6a CLIMOGRAPH—RANGOON, MYANMAR

Climate	Tropical Monsoon
Annual Temperature Range	5.5°C
Avg. Annual Temperature	27.3°C
Summer Precipitation	2265.0 mm
Winter Precipitation	264.0 mm
Total Annual Precipitation	2529.0 mm

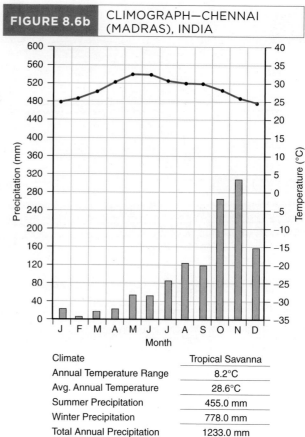

FIGURE 8.6b CLIMOGRAPH—CHENNAI (MADRAS), INDIA

Climate	Tropical Savanna
Annual Temperature Range	8.2°C
Avg. Annual Temperature	28.6°C
Summer Precipitation	455.0 mm
Winter Precipitation	778.0 mm
Total Annual Precipitation	1233.0 mm

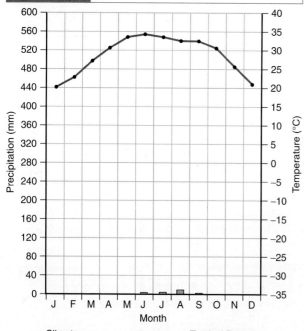

FIGURE 8.6c CLIMOGRAPH—FAYA, CHAD

Climate	Tropical Desert
Annual Temperature Range	13.8°C
Avg. Annual Temperature	28.7°C
Summer Precipitation	16.0 mm
Winter Precipitation	0.0 mm
Total Annual Precipitation	16.0 mm

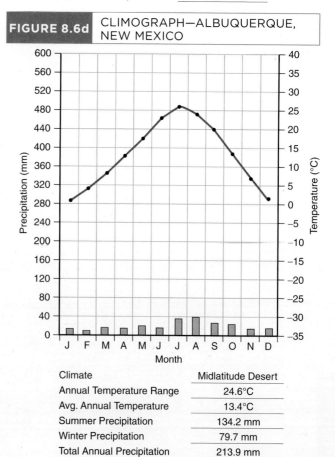

FIGURE 8.6d CLIMOGRAPH—ALBUQUERQUE, NEW MEXICO

Climate	Midlatitude Desert
Annual Temperature Range	24.6°C
Avg. Annual Temperature	13.4°C
Summer Precipitation	134.2 mm
Winter Precipitation	79.7 mm
Total Annual Precipitation	213.9 mm

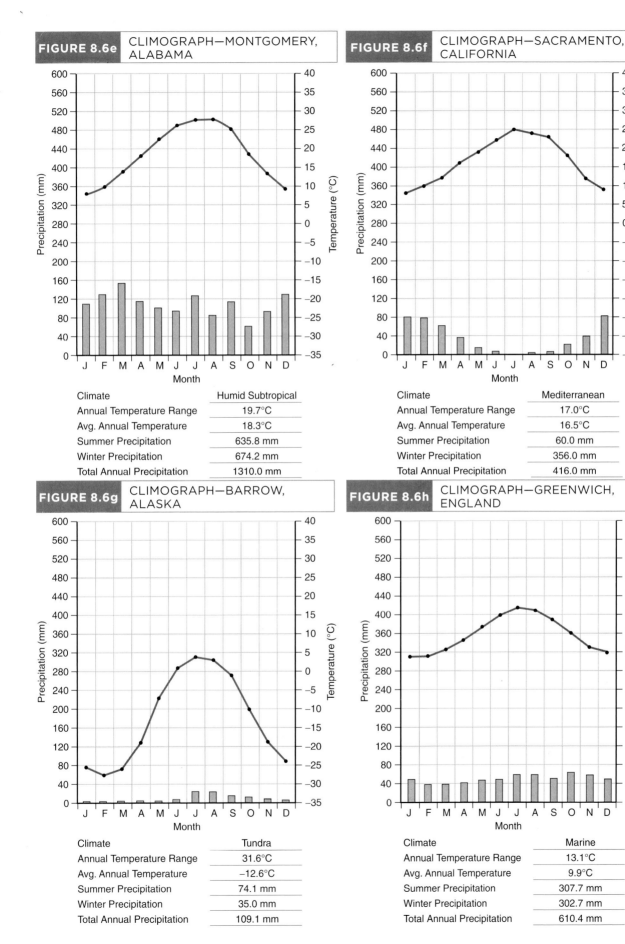

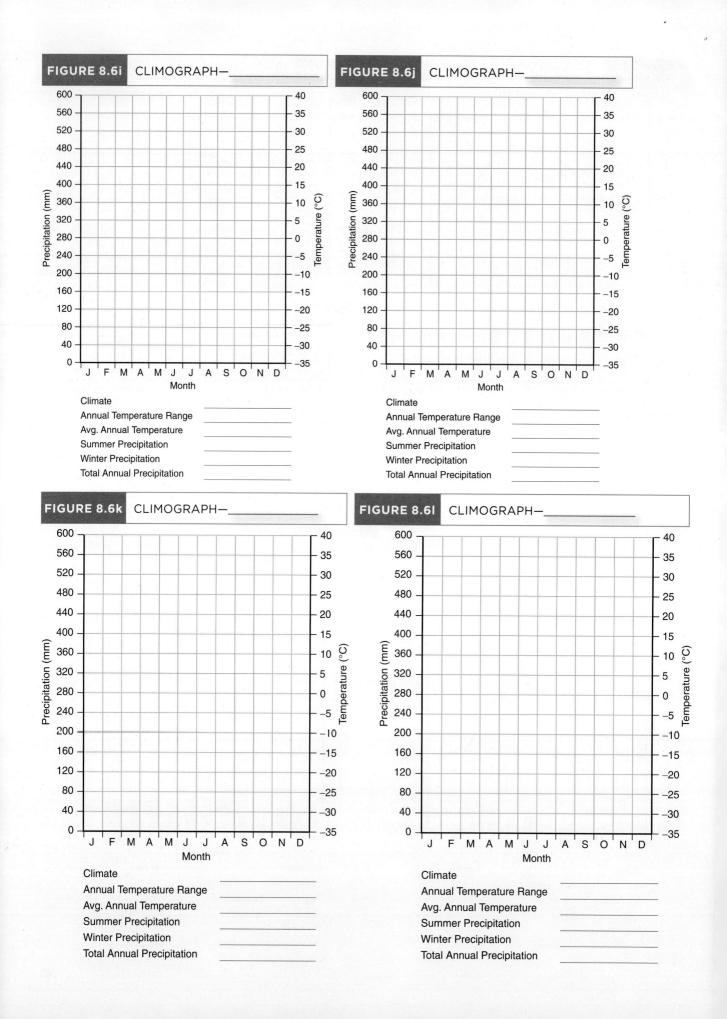

4. Use Tables 8.2 and 8.3 and Figures 8.1 through 8.4 to classify each station's climate.
   Use the data in Table 8.3, the summary data associated with the climographs in Figure 8.6, and your knowledge of climates to answer the remaining questions.

5. a. During which sun period (high or low) does most of the precipitation occur in Iquitos, Peru?

   b. What is responsible for the large amount of precipitation during this time?

6. Compare and contrast the seasonal temperature and precipitation variations of the rainforest and monsoon climates.

7. What is responsible for Colorado Springs' dry climate?

8. How do the midlatitude desert (BWk) climates differ from the tropical desert (BWh) climates?

9. Sacramento, California, and Montgomery, Alabama, are located at about the same latitude and have similar average annual temperatures. Why do they have such different precipitation regimes?

10. a. When (summer or winter season) does most of the precipitation fall in the subarctic climates?

    b. Why?

11. Why does Sacramento receive most of its precipitation during the winter?

12. a. Which climate station has the largest temperature range? _____

    b. What is responsible for its large temperature range?

13. a. On the graph provided, plot the temperature range (listed in Figure 8.6) on the *Y*-axis versus latitude (listed in Table 8.3) on the *X*-axis for each of the stations in Table 8.3. There is no need to differentiate between Northern and Southern Hemisphere locations.

    b. Estimate the position and draw a trend line (straight line) through the plotted points.

14. Describe the relationship between latitude and temperature range represented by your trend line.

15. a. Which points seem to deviate vertically the most from the trend line?

    b. Is there anything about their location that explains their large deviations?

Name: Dulce Lizarraga   Section:

18.55 / 20

PART 2 • INFLUENCE OF OCEAN CURRENTS ON COASTAL CLIMATE

To-Do bx 4/30/18

The five predominant subtropical high pressure cells are shown in Figure 8.7.

1. Draw the location of following ocean currents on Figure 8.7 using color coded arrows, blue for cold currents and red for warm. Consult an atlas or your textbook.

   - ✓ Gulf Stream Current (warm)
   - ✓ North Atlantic Current (warm)
   - ✓ California Current (cold)
   - ✓ Brazil Current (warm)
   - ✓ Peru Current (cold)
   - ✓ Canary Current (cold)
   - ✓ Benguela Current (cold)
   - ✓ Kuroshio Current (warm)
   - ✓ North Equatorial Current (warm)
   - ✓ South Equatorial Current (warm)

2. Draw arrows showing the air circulation around the subtropical highs on Figure 8.7.

3. a. Is there a relationship between the location of a particular type of ocean current (warm or cold) and coast (east or west), especially in the subtropics and midlatitudes?

   Yes,

   b. If so, what is it?

   The North Equatorial currents and the South Equatorial currents. The West coast have colder temperatures as compare to the warm temperatures of the east coast.

   c. Speculate on the reason for the relationship.

   The warm equatorial currents meet, coming from the North & south Westerlies cold currents.

   Coriolis

   gyres

   - Corealis causes the gyres to rotate clockwise on the Norther hemisphere and counter clockwise in the southern hemisphere

4. Plot and label the location of San Diego, California (32°N, 117°W), and Jacksonville, Florida (30°N, 82°W), on Figure 8.7.

Orographic Effect — Rain Shadow — Ocean — rain — Dry

+2

FIGURE 8.7 THE FIVE PREDOMINANT SUBTROPICAL HIGH PRESSURE CELLS

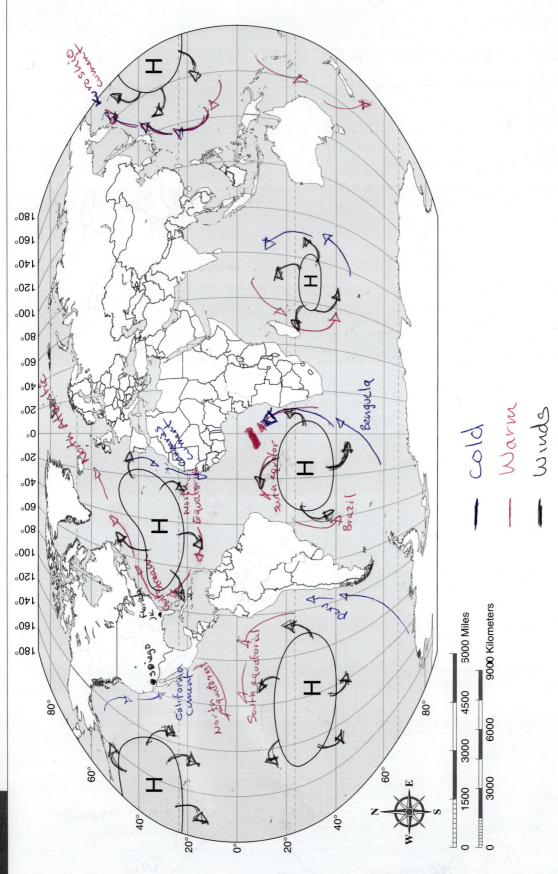

5. Use the data in Table 8.4 to calculate the statistics for Table 8.5. Be sure to show your units.

TABLE 8.4	CLIMATE DATA FOR TWO CITIES											
	Jan	Feb	Mar	Apr	May	Jun	Jul	Aug	Sep	Oct	Nov	Dec
San Diego, California (32°N, 117°W)												
T (°C)	12.7	13.3	14.2	15.5	16.8	18.4	20.4	21.2	20.3	18.2	15.6	16.7
P (mm)	47.2	47.5	39.9	18.3	6.8	1.4	0.8	2.2	3.4	10.7	25.5	46.7
Jacksonville, Florida (30°N, 82°W)												
T (°C)	12.9	14.0	17.0	20.3	23.8	26.6	27.7	27.5	25.8	21.4	16.8	13.4
P (mm)	72.4	81.4	86.0	70.3	96.3	145.7	168.2	168.1	182.9	110.4	50.0	69.9

Annotations: 203.3/12, 250.4; 247.20/12, 1,301.6; "evaporation →" by T row, "precipitation →" by P row

TABLE 8.5	SUMMARY CLIMATOLOGICAL STATISTICS	
	San Diego	**Jacksonville**
Mean annual temperature (Average)	16.94 °C	20.6 °C
Annual temperature range	8.5 °C	14.8 °C
Total annual precipitation	250.4 mm	1,301.6 mm

6. What kind of ocean current (warm/cold) flows along the coast of California near San Diego?

   Cold

7. What effect does the ocean current have on air temperature and stability at San Diego?

   Keeps the air stable, and it keeps the temperature steadier

8. How does the stability of the air help to explain the dry conditions at this coastal location?

   at 70°F it keeps the air stable, and it keeps the temperature stable. The stability of the air keeps the temperature dry, and the air does not rise

9. What kind of ocean current (warm/cold) is found off the coast of Florida near Jacksonville?

   Warm currents

10. What effect does the ocean current have on the temperature and stability of air at Jacksonville?

✓ Ustable air, lot more rainful, it creates a lot of temperature range for the clouds to form

11. How does the stability of the air influence the precipitation at Jacksonville?

✓ the low stability in Jacksonville helps the clouds to form when it arises to create rain.

Name: _____   Section: _____

PART 3 · OROGRAPHIC INFLUENCE ON CLIMATE

To-Do by 4/30/18

1. Label the following mountain ranges on the world map in Figure 8.8. You will have to consult an atlas.

   ✓ Rocky Mountains   ✓ Sierra Nevada   ✓ Caucuses
   ✓ Andes   ✓ Appalachian   ✓ Himalayas
   ✓ Cascades   ✓ Alps   Hindu Kush

2. Plot and label the locations in Table 8.6 on the world map in Figure 8.8.

TABLE 8.6	CLIMATE DATA FOR SELECTED CITIES											
	Jan	Feb	Mar	Apr	May	Jun	Jul	Aug	Sep	Oct	Nov	Dec
Brest, France (49°N, 4°W)												
T (°C)	6.2	6.0	7.4	8.8	11.5	14	15.8	15.9	14.6	12.1	8.8	7.4
P (mm)	131.8	105.6	101.4	72.7	72.1	57.9	49.1	68.3	85.2	110	127.1	147.9
Strasbourg, France (49°N, 8°E)												
T (°C)	0.0	1.9	5.2	9.5	14	17.2	18.8	18.1	14.7	9.7	4.7	1.4
P (mm)	34.7	31.8	36.7	46.1	65.5	72.8	75.2	70.4	60.3	49.8	46.9	39.2
Orleans, France (48°N, 2°E)												
T (°C)	3.4	3.5	6.9	9.3	13.2	16.3	19.1	18.4	16.1	12.3	7.0	4.7
P (mm)	49.1	40.8	45.2	48.1	57.3	55.0	53.8	52.6	50.0	62.3	58.8	53.6
Astoria, Oregon (46°N, 124°W)												
T (°C)	5.5	6.7	7.6	9.0	11.3	13.8	15.5	15.9	14.6	11.4	8.2	5.7
P (mm)	256.0	197.4	179.6	124.2	72.4	64.3	27.6	34.4	70.7	154.7	264.2	266.0
Walla Walla, Washington (46°N, 118°W)												
T (°C)	−2.1	0.9	4.7	9.9	14.7	19.4	24.7	23.6	18.3	11.5	3.4	−0.2
P (mm)	34	30	40	45	36	25	15	22	13	29	33	31
Butte, Montana (49°N, 113°W)												
T (°C)	−8.4	−5.5	−2.0	3.4	8.4	13.3	17.1	16.2	10.4	5.0	−2.3	−8.0
P (mm)	14.9	12.1	19.5	25.8	46.7	60.8	32.5	30.1	27.6	19.7	13.9	13.9

FIGURE 8.8 WORLD MAP WITH MOUNTAIN RANGES

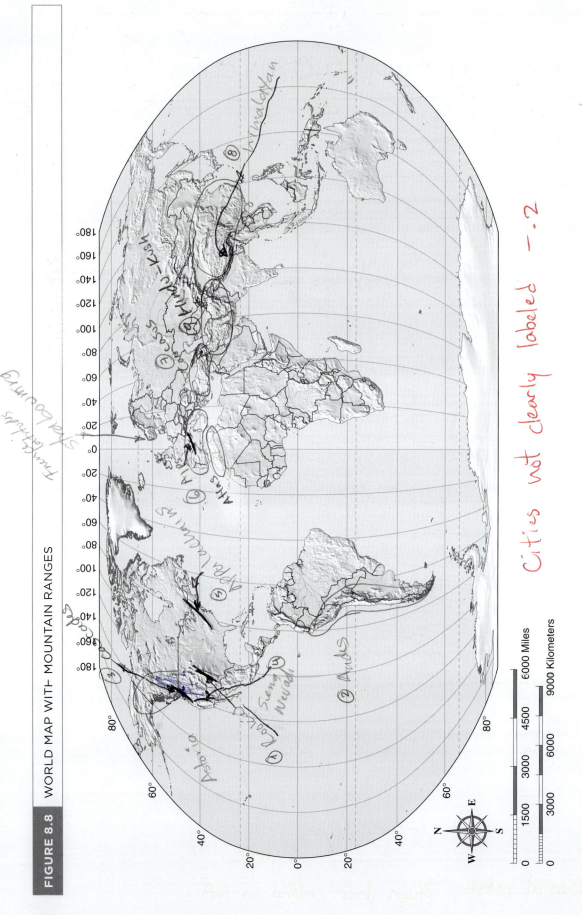

3. Examine your world map, then describe the general orientation of mountains in North America and Europe.

   North america mountains go down south, compare to the european mountains from west to east. ✓

4. Which air mass likely dominates Astoria, Oregon, on the west coast of North America and Brest, France, on the west coast of Europe? Maritime Polar air mass ✓

5. What is the prevailing wind direction and air mass movement at the latitudes for Astoria and Brest?

   From west to east by the westerlies. ✓

6. a. Plot the total annual precipitation for Astoria, Walla Walla, and Butte on the graph provided in Figure 8.9 and connect the points with straight-line segments. Indicate the relative location of any mountains between Astoria and Butte on the graph. Consult your world map (Figure 8.8).

   b. Plot the total annual precipitation for Brest, Orleans, and Strasbourg on the graph provided in Figure 8.9 and connect the points with straight-line segments. Indicate the relative location of any mountains between Brest and Strasbourg on the graph. Consult your world map (Figure 8.8).

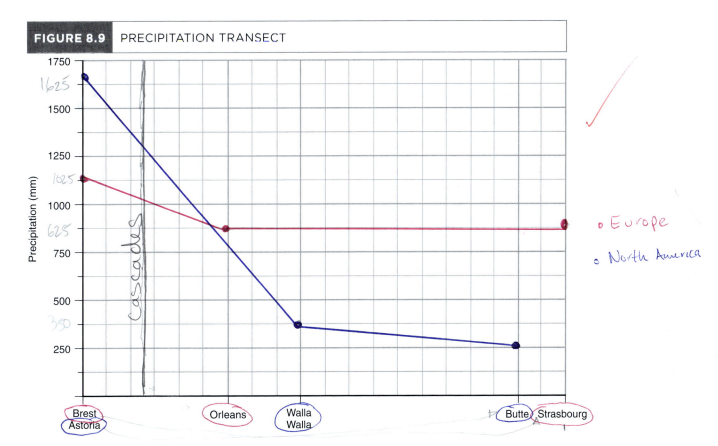

FIGURE 8.9 PRECIPITATION TRANSECT

7. Compare and contrast the west-to-east pattern of precipitation from Astoria to Butte with the pattern of precipitation from Brest to Strasbourg.

   With the Orographic effect coming from Astoria (west coast), the mountain stops the precipitation forming from the ocean to a Dry Rain Shadow. as soon the rain passed the mountains. true...

   high precipitation at coast, much drier inland in NA

   −.25

   +5.75

8. Astoria and Brest are both located very close to a coast yet Astoria receives much more precipitation. Why?

   because of the influence ~~of polar air mass~~

   because of orographic effect caused by Cascades

   −1

9. The annual temperature range for Strasbourg is 18,8°C while Butte has an annual temperature range of 25°C. Why is the temperature range so much smaller in Strasbourg than in Butte even though they are nearly the same distance from the ocean? because of the orientations of the mountains Strasbourg mountains are sideways and do not block as much air to stop the precipitation. So tends to be cooler.

10. Explain how mountain orientation and air mass movement affect the spatial distribution of precipitation on the west coast climate of North America and Europe based on your answers above, the world map in Figure 8.8, and a global map of climate. The way the mountains are distributed; North A. from North to South and Europe are destributed from west to East.

94

+0

EXERCISE 1 ⎮ EARTH-SUN GEOMETRY AND INSOLATION

PURPOSE

The purpose of this exercise is to learn relationships between the earth and the sun, and to examine how spatial and temporal variations in sunlight affect temperature and energy patterns on the earth's surface.

LEARNING OBJECTIVES

By the end of this exercise you should be able to

- calculate the noon sun angle of a place during the solstices and equinoxes;
- explain how sun angle and insolation intensity changes with latitude and season;
- explain how day length changes with latitude and season; and
- explain how sun angle, insolation intensity, and day length affect temperature.

INTRODUCTION

The earth's axis is tilted 23.5° away from perpendicular to the plane of the earth's orbit around the sun. The **tilt of the earth's axis** produces seasonal changes in the angle at which the sun's rays strike any given location on the earth, in the length of day, and in the amount of energy reaching any given location (Figure 1.1).

The amount of solar radiation reaching the outside of the earth's atmosphere remains fairly constant. Averaging the effects of the earth's elliptical orbit around the sun, the earth receives approximately 1367.7 watts per square meter (1367.7 W/m^2), or 2 calories per square centimeter per minute (2 cal/cm^2/min) on a plane-oriented perpendicular to the sun at the outer edge of the atmosphere, a value known as the **solar constant.**

The solar constant represents the maximum energy available to the earth-atmosphere system. However, radiation is unequally distributed across the face of the earth and is reduced as it travels through the atmosphere. Several factors determine the total amount of incoming solar radiation, or **insolation**, that strikes the earth's surface:

1. **Sun angle**—the angle a beam of light makes with the surface of the earth. Low latitudes receive more insolation than high latitudes because sun angles are greater in the tropics.
2. **Day length period**—the longer the sun remains above the horizon, the more total insolation any given location will receive.
3. Water vapor content and cloud coverage of the atmosphere—water, regardless of its state (liquid, solid, or gas), absorbs and reflects solar radiation. Therefore, less radiation is received at the earth's surface on humid or cloudy days than on dry or clear days.
4. **Atmospheric path length** or thickness—the amount of atmosphere through which the sun's rays pass before striking the earth's surface. High mountains receive more solar radiation than lower elevations because the atmosphere is thinner at high elevations. Likewise, the solar beam must traverse through more atmosphere when it is on the horizon than when it is directly overhead. The greater the path length, the greater the opportunity for absorption or reflection and the lower the insolation.

This exercise examines the first two factors, sun angle and day length period.

Earth-Sun Geometry and Sun Angle

The **sun angle** is the angle that a beam of light makes with respect to the earth's surface. In other words, it is how high the sun appears above the horizon. The latitude of a place, the time of year, and the time of day largely determine the sun angle. The sun angle increases during the morning reaching a peak at noon and then decreases into the afternoon. The **noon sun angle** (S), is the angle of the sun above the horizon at noon. Barring the influence of clouds and other factors that would deplete the incoming light, the most intense radiation occurs at noon when the sun is highest in the sky.

The **zenith angle** (Z) is the angle formed between a line perpendicular to the earth's surface and the position of the sun in the sky (Figure 1.2). If the sun is on the horizon at noon, the noon sun angle (S) is 0° and the zenith angle (Z) is 90°; if the sun is directly overhead at noon, the noon sun angle (S) is 90° and the zenith angle (Z) is 0°. The zenith angle for a selected location can be determined by counting the degrees of latitude that separate the selected location from the declination of the sun.

FIGURE 1.1	TEMPORAL VARIATIONS IN SUN ANGLE AND DAYLIGHT

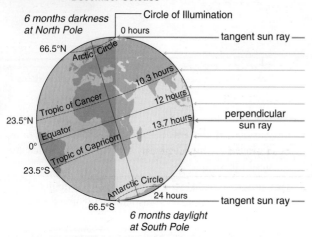

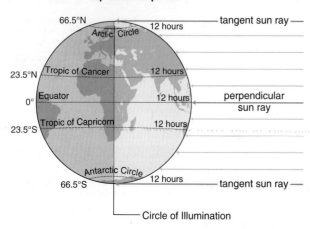

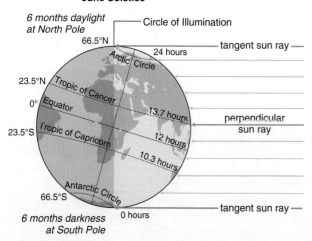

The **declination** of the sun is the latitude where the sun on any day is directly overhead (S = 90°) at noon; thus, the declination is determined by the date. The declination of the sun varies from 23.5°N on the **June solstice** to 23.5°S on the **December solstice.** Between these latitudes (23.5°N to 23.5°S), the sun passes directly overhead twice during the

FIGURE 1.2	ANNUAL CHANGES IN THE NOON SUN ANGLE FOR 50°N LATITUDE

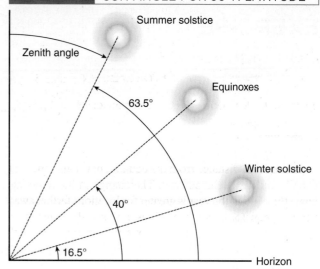

year. For example, at the equator (latitude = 0°) the noon sun angle is 90° on the **March and September equinoxes.** Outside these latitudes (23.5°N to 23.5°S), the sun is never directly overhead. The noon sun angle, S, always equals 90° − Z. Thus, there are four steps for calculating the noon sun angle for any selected location (Table 1.1):

1. Determine the <u>latitude</u> of the selected location.
2. Determine the declination of the sun, which is determined by the date.
3. Calculate the zenith angle, Z.
4. Calculate the noon sun angle, S.

TABLE 1.1	EXAMPLES OF COMPUTING THE NOON SUN ANGLE

1. Noon sun angle for 40°N latitude on June 21:
   Declination = 23.5°N
   Z = 40°N − 23.5°N = 16.5°
   S = 90° − 16.5° = 73.5°
   For this example, at noon the sun would appear to the south of the observer.

2. Noon sun angle for 40° N latitude on December 22:
   Declination = 23.5°S
   Z = 40°N + 23.5°S = 63.5°
   S = 90° − 63.5° = 26.5°
   In this example, at noon the sun would appear to the south of the observer.

3. Noon sun angle for 5°S latitude on June 21:
   Declination = 23.5°N
   Z = 5°S + 23.5°N = 28.5°
   S = 90° − 28.5° = 61.5°
   In this example, the sun would appear to the north of the observer because the declination is located to the north of the observer.

The direction that the sun appears relative to an observer depends on two things: (1) the observer's latitude, and (2) the declination of the sun. If the observer's location (latitude) is outside the latitude zone between 23.5°N and 23.5°S, the sun always appears to the south of an observer located in the northern hemisphere, and to the north of an observer located in the southern hemisphere. Between 23.5°N and 23.5°S, the sun's location in the sky can vary. For instance, on the March equinox, the declination is 0°. To a person located at 10°N, the sun would appear in the southern portion of the sky. However, on the June solstice when the declination is 23.5°N, the sun would appear in the northern portion of the sky.

Latitude, or distance from the equator, plays an important role in earth-sun relationships. The change in the noon sun angle throughout the year is greater for locations farther away from the equator than for locations closer to the equator, up to a point. This is true whether one moves north or south of the equator. For example, at the equator, the noon sun angle ranges from a high of 90° on the equinoxes to a low of 66.5° on the solstices, a difference of 23.5°. As shown in Figure 1.2, the noon sun angle at 50°N latitude ranges from a high of 63.5° on the June solstice to a low of 16.5° on the December solstice, a difference of 47°. At the North Pole, 90°N, the noon sun angle ranges from a high of 23.5° to a low of 0°, a difference of 23.5°. These changes in noon sun angle throughout the year affect temperature changes throughout the year. In addition, the angle itself affects the intensity of sunlight and thus the temperature.

The angle at which the sun's rays strike the earth's surface will determine the amount of area those rays will cover. As the angle of the sun decreases, the area covered increases. This means that the amount of energy received at a specific point will be less simply because the energy is spread out over a greater area (Figure 1.3).

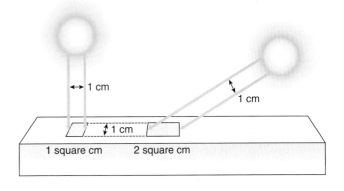

FIGURE 1.3 AREA ILLUMINATED BY THE SUN

The intensity of radiation falling on a surface varies with the sine of the sun angle. When sun angles are high, solar energy is concentrated in a smaller area and the intensity of energy input is high. When sun angles are low, the energy is spread over a larger area, reducing the intensity of energy input. Table 1.2 gives the sun's intensity for a variety of sun angles as a percentage of a perpendicular beam. Tens of degrees of sun angle (i.e., 0 – 80 degrees) run down the column on the far left-hand side of the table. Single units (0 – 9 degrees) run along the top row of the table. To find the intensity of any given sun angle, such as a noon sun angle of 74°, read down the column of tens to the number 70 and across the row until you intersect the unit column of 4. Thus a noon sun angle of 74° would have a radiation intensity of 96.13%. This means that the radiation intensity of this sun angle would be about 96% of a perpendicular (90°) angle. Note that the highest intensity will be for an angle of 90°

TABLE 1.2 INTENSITY OF SOLAR RADIATION*

	Degrees of Sun Angle	Units									
		0	1	2	3	4	5	6	7	8	9
	0	00.00	01.75	03.49	05.23	06.98	09.72	10.45	12.19	13.92	15.64
	10	17.36	19.08	20.79	22.50	24.19	25.88	27.56	29.24	30.90	32.56
	20	34.20	35.84	37.46	39.07	40.67	42.26	43.84	45.40	46.95	48.48
	30	50.00	51.50	52.99	54.46	55.92	57.36	58.78	60.18	61.57	62.93
	40	64.28	65.61	66.91	68.20	69.47	70.71	71.93	73.14	74.31	75.47
	50	76.60	77.71	78.80	79.86	80.90	81.92	82.90	83.87	84.80	85.72
Tens	60	86.60	87.46	88.29	89.10	89.88	90.63	91.36	92.05	92.72	93.36
	70	93.97	94.55	95.11	95.63	96.13	96.59	97.03	97.44	97.81	98.16
	80	98.48	98.77	99.03	99.25	99.45	99.62	99.76	99.86	99.94	99.98

* Radiation intensity (%) for given sun angles—expressed as a percentage of the radiation intensity of a perpendicular beam (90 degrees).

(a perpendicular beam) and the lowest for an angle of 0° (one right on the observer's horizon).

Day Length Period

Day length period also changes throughout the year. Changes in day length are greatest at the North and South Poles, ranging from zero hours of sunlight on the winter solstice to 24 hours of sunlight on the summer solstice. Changes in daylight hours are least at the equator. There are 12 hours of daylight and dark at the equator every day of the year. Table 1.3 lists the hours of daylight for different latitudes on the December solstice.

IMPORTANT TERMS, PHRASES, AND CONCEPTS

tilt of the earth's axis
solar constant
insolation
sun angle
day length period
atmospheric path length (thickness)
noon sun angle
zenith angle
declination
June and December solstices
March and September equinoxes

TABLE 1.3 LENGTH OF DAYLIGHT DURING THE DECEMBER SOLSTICE

Latitude	Daylight (hr:min)	Latitude	Daylight (hr:min)
90°N	0:00	0°	12:07
80°N	0:00	10°S	12:28
70°N	0:00	20°S	13:05
60°N	5:52	30°S	13:48
50°N	8:04	40°S	14:40
40°N	9:20	50°S	15:56
30°N	10:12	60°S	18:08
20°N	10:55	70°S	24:00
10°N	11:32	80°S	24:00
0°	12:07	90°S	24:00

Name: Dulce Marlizarraga-Chagolla   Section: _____

17.25 / 20

EARTH-SUN GEOMETRY AND INSOLATION

1. Calculate the declination, zenith angle, noon sun angle and radiation intensity (see Table 1.2) for Minneapolis, Minnesota and Belize City, Belize for each date.
2. Plot the values of the noon sun angle for Minneapolis and Belize City (Tables 1.4 and 1.5) and connect the points with short line segments on the graph provided in Figure 1.4.

TABLE 1.4 EARTH-SUN GEOMETRY AND INSOLATION FOR MINNEAPOLIS, MINNESOTA, 45°N

Date	Declination	Zenith angle	Noon sun angle	Radiation intensity (%)
March 21 20th	0°	45°	45°	70.71 %
June 21	23.5°N	21.5°	68.5°	93.04 %
September 23	0°	45°	45°	70.71 %
December 22	23.5°S	45° 68.5°	21.5°	36.65 %

Cuando es del mismo lado Norte/norte / Sur/sur ✓

TABLE 1.5 EARTH-SUN GEOMETRY AND INSOLATION FOR BELIZE CITY, BELIZE, 17.5°N

Date	Declination	Zenith angle	Noon sun angle	Radiation intensity (%)
March 21	0°	17.5°	(90-17.5) = 72.5°	95.37 %
June 21	23.5°N	6°	(90-6) 84°	99.45 %
September 23	0°	17.5°	(90-17.5) = 72.5	95.37 %
December 22	23.5°S	41°	(90-41) = 49°	75.47 %

FIGURE 1.4 VARIATION IN NOON SUN ANGLE

• Belize ✓

• Minneapolis

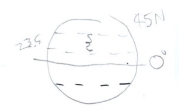

3. Which of the two locations on page 5, Minneapolis or Belize City, experiences the greatest change in noon sun angle throughout the year?

   Minneapolis ✓

4. Based on the information for Minneapolis and Belize City in Tables 1.4 and 1.5, what effect does the sun angle have on radiation intensity at the earth's surface? *higher sun angle = higher radiation intensity* −.25

   Sun rays are more prominent in Belize than Minneapolis, meaning that in Belize tends to be hotter (tropical weather) radiation intensity is higher.

5. What effect might the amount of change in <u>noon sun angle</u> at these two cities have on their respective annual <u>temperature ranges</u> (<u>difference</u> between <u>maximum</u> monthly temperature and <u>minimum</u> monthly temperature)?

   The maximum solar intensity in Minneapolis is about 93% while in Belize is about 99% in summer time, which Belize temperature are higher than Minneapolis in summer. On winter time is the opposite, Minneapolis temperature drops to the single degrees, while on belize stays up with high radiation intensity. −.5

6. a. Would the sun appear to the north or south of an observer in each of these cities on June 21?

   Minneapolis: South ✓

   Belize City: North

   b. Where would the sun appear on December 22?

   Minneapolis: South ✓

   Belize City: South

   c. Why does the sun appear in different directions at these two cities?

   Because the sun never goes directly outside the tropic lines, and in at different latitudes −1

7. Fill in the information for the following locations on December 22. Use Table 1.3 to estimate the day length period. (Note: Location latitudes have been rounded to nearest half degree.)

8. a. Plot the values of the noon sun angle (Y-axis) and latitude (X-axis) from Table 1.6 on the graph paper provided and connect the points with short line segments.

   b. Plot the values of daylight (Y-axis) and latitude (X-axis) from Table 1.6 on the graph paper provided and connect the points with short line segments.

9. Using your answers to questions 7 and 8 as an example:

   a. How would the number of hours of daylight change as you travel from the North Pole to the South Pole on the <u>June</u> solstice?

   It gradually changes from the south to North with more sunlight until it reach up to 24 hrs of daylight.

   more detail −.25

6

+5

23.5 South
Dec Solstice

Location	Latitude	Zenith angle	Noon sun angle	Daylight (hr:min)
Murmansk, Russia	69.0°N +23.5° South	92.5°	−2.5°	◯
Krakow, Poland	50.0°N +23.5°S	73.5°	16.5	8:04
Nicosia, Cyprus	35.0°N +23.5°S	58.5°	(90−58.5)=31.5°	9:20+26m.= 9:46m
Nasser Lake, Egypt	24.0°N +23.5°S	47.5°	(90−47.5)=42.5°	10:55−19=10:36
Nairobi, Kenya	23.5°S − 1.5°S	22°	(90−22°)=68°	12:07+.02=12:09 hrs min
Noumea, New Caledonia	23.5°S − 22.5°S	1°	(90−1°)=89°	13:16
Wellington, New Zealand	41.5°S −23.5°S	18°	(90−18°)=72°	14:45
Ross Ice Shelf, Antarctica	81.5°S −23.5°S	58°	(90−58°)=32°	24 hrs.

b. How would the noon sun angle change as you travel from the North Pole to the South Pole on the June solstice?

It would increase until I reach the tropic of cancer and as I go down to ___ and pass it would decrease down to 0° to the South Pole.

c. On the June solstice, at what latitude (approximately) would you expect to experience the warmest temperatures, and at what latitude (approximately) would you expect to experience the coldest temperatures?

warmest: at the tropic cancer at 23.5°

coldest: at the South Pole at 0° 90°S −.5

10. In which of the following cases will a beam of solar radiation be dispersed over the *smallest* surface area when it strikes the earth's surface at the place and time indicated?

  0° a. Cape Town, South Africa, at noon on March 21
  23.5°S b. Singapore, Malay Peninsula, at noon on December 21
  23.5°N c. Kolkata (Calcutta), India, at noon on June 21 this one
  0° d. Buenos Aires, Argentina, at noon on March 21
  0° e. New Orleans, Louisiana, at noon on September 21

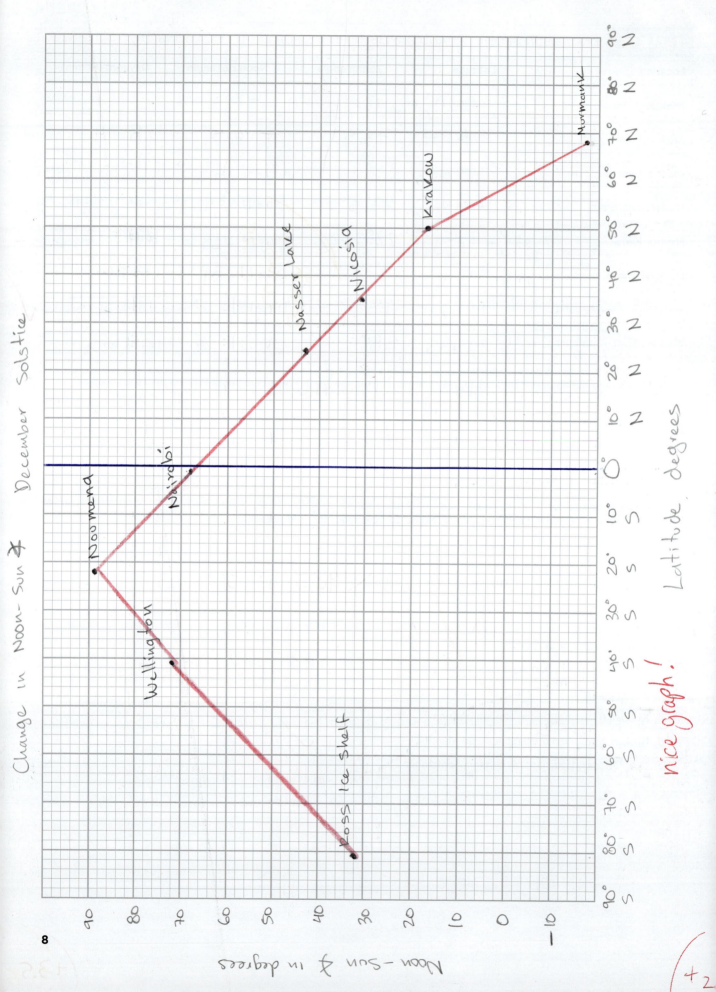

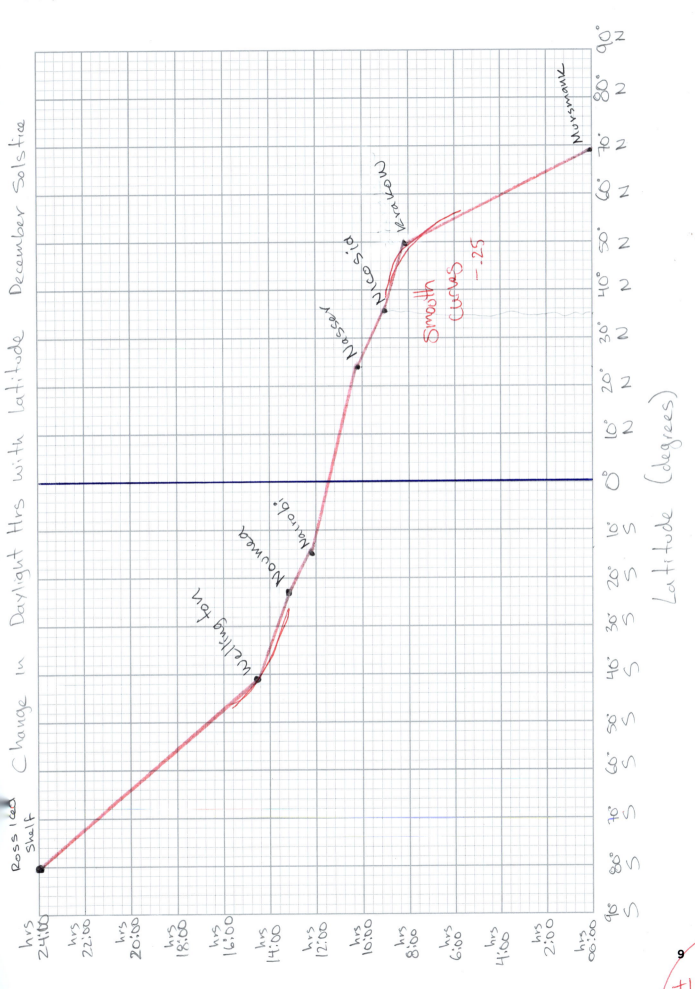

EXERCISE 3 ATMOSPHERIC TEMPERATURE

PURPOSE

The purpose of this exercise is to develop an understanding of the processes and factors that control the spatial variation of air temperature.

LEARNING OBJECTIVES

By the end of this exercise you should be able to

- convert between temperature scales;
- explain the effect of continentality on air temperature; and
- calculate a latitudinal temperature gradient and predict seasonal temperatures.

INTRODUCTION

Temperature is a measure of the average kinetic energy of atoms or molecules comprising a substance. Temperature is the hotness or coldness of a substance. **Heat** is the total energy associated with random atomic and molecular motions of a substance. Heat represents the thermal energy that flows from a body of higher temperature to a body of lower temperature. The total energy is dependent on the mass and the temperature of a body. Many factors control the temperature of the air and ground surface. Location (latitude in particular) and atmospheric conditions determine the amount and intensity of radiation that strikes the surface of the earth. Properties such as **specific heat** determine how much energy is required to raise the temperature of different substances. For instance, the specific heat of water is approximately five times greater than that of land. Given equal amounts of energy input, it will take much longer for water to heat up (and give off heat) than land. The temperature of the air above land and water surfaces will reflect such properties.

The movement of air masses in association with weather fronts or pressure systems can significantly alter the air temperature at a place. The penetration of Canadian arctic air behind a fast-moving cold front into the midlatitudes of North America may cause an extreme drop in temperatures over a short period of time. However, the land surface can act to modify moving air masses as well. During the winter, the temperature of air masses that originate over the relatively warm ocean will decrease in temperature when traveling over cold continental surfaces. During the summer, cool oceanic air masses gain energy from the warm continental land surface as they march into the interior.

Temperature Scales and Statistics

There are two temperature scales used in this lab, the **Celsius** (°C) and **Fahrenheit** (°F) scales. Scales are based on the physical properties of substances such as water. For instance, on the Celsius scale, water freezes at 0°C (32°F) and boils at 100°C (212°F). The formulas for converting between scales are:

$$1.8°C + 32 = °F \tag{3.1}$$

$$(°F - 32)/1.8 = °C \tag{3.2}$$

A variety of temperature statistics are compiled by weather services. These statistics aid weather forecasts and assessment of climatic patterns and trends. Some of the more common temperature statistics are:

$$\text{Average annual temperature} = \Sigma \text{ MMT}/12 \tag{3.3}$$
where MMT = mean monthly temperature

$$\text{Annual temperature range} = \text{warmest MMT} - \text{coldest MMT} \tag{3.4}$$

$$\text{Maximum range in annual temperature} = \text{highest maximum monthly temperature} - \text{lowest minimum monthly temperature} \tag{3.5}$$

Continentality and Air Temperature

Continentality is the effect of location within a continental landmass on climate. Places located in the interior of a continent tend to have large temperature ranges and shorter lag periods between maximum incoming radiation and maximum temperature. This is due to the heating characteristics of land surfaces versus those of water, as we discussed at the beginning of this exercise. Air above a water surface will heat up more slowly (and cool down more slowly) than air above a land surface. Also, the air above water will not attain as high a temperature as that above land. In addition, the transparency of water allows sunlight to penetrate through it, and water has the ability to mix, thus redistributing its heat. Land cannot do this. Because of these properties, large bodies of water are often called "heat reservoirs." Ocean currents act to redistribute heat gained in regions of energy surplus to regions of energy deficits. Warm ocean currents transfer energy from equatorial regions toward the Poles, while cold ocean currents move colder water toward the equator.

Latitudinal Temperature Gradients

The atmosphere above the earth's surface is heated primarily by the absorption of longwave radiation emitted by the earth and by the conduction of energy between the earth's surface and the air above. This energy is derived from the absorption of solar radiation by the earth. Primarily earth-sun geometry and latitude control the amount of energy received from the sun. The general decrease in receipt of solar radiation at the earth's surface with increasing latitude creates a north-south air temperature gradient, which has a significant impact on the distribution of air pressure, winds, and weather systems around the earth.

Graphs of air temperature versus latitude may reveal spatial variations in temperature, as shown in Figure 3.1. In addition, temperature gradients may be calculated from these graphs. In Figure 3.1, the points plotted on the graph represent temperatures at different latitudes. After plotting the points, a trend line was drawn through the middle of the points to show the trend of temperature versus latitude.

A **gradient** is defined as the rate of change in a variable (in this case, temperature) through space, either vertical space (altitude) or horizontal (across the earth's surface). The **latitudinal temperature gradient** is:

$$\text{Latitudinal temperature gradient} = \frac{\Delta \text{Temperature}}{\text{Distance}} \quad (3.6)$$

where the Greek letter delta (Δ) stands for "change in." To calculate the latitudinal temperature gradient, two points on the trend line were selected, (X_1, Y_1; X_2, Y_2). The gradient is calculated as:

$$\text{Latitudinal temperature gradient} = \frac{(Y_2 - Y_1)}{(X_2 - X_1)} = \frac{(\text{Temperature}_2 - \text{Temperature}_1)}{(\text{Latitude}_2 - \text{Latitude}_1)}$$

Substituting in the values of:

$$(X_1, Y_1) = (20°\text{Lat}, 25°\text{C})$$
$$(X_2, Y_2) = (50°\text{Lat}, 5°\text{C})$$

we get:

$$\text{Latitudinal temperature gradient} = \frac{(5°C - 25°C)}{(50°\text{Lat} - 20°\text{Lat})} = \frac{-20°C}{30°\text{Lat}} = \frac{-0.67°C}{1°\text{Lat}}$$

Using these two points, the temperature gradient equals $-0.67°C$ per $1°$ of latitude. This simply means that for every $1°$ increase in latitude, the temperature will decrease by $0.67°C$. Thus we have found a way to characterize the spatial variation of temperature near the surface.

IMPORTANT TERMS, PHRASES, AND CONCEPTS

temperature
continentality
Celsius
Fahrenheit

heat
specific heat
latitudinal temperature gradient
gradient

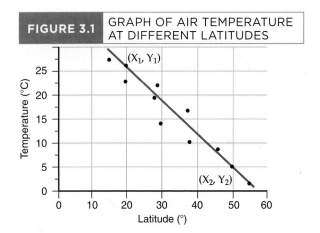

FIGURE 3.1 GRAPH OF AIR TEMPERATURE AT DIFFERENT LATITUDES

$(1.8 \times C°) + 32 = F$

Name: _____ **Section:** _____

$1.8(°C) + 32 = °F$ $\dfrac{(°F - 32)}{1.8} = °C$

PART 1 · TEMPERATURE CONVERSIONS

1. Convert the following Fahrenheit temperatures to Celsius.
   a. 32°F   = __14.22__ °C $32°F - 32 \div 1.8 = 14.22°C$
   b. 50°F   = __32.22__ °C $50°F - 32 \div 1.8 = 32.22°C$
   c. −22°F = __−39.77__ °C $-22°F - 32 \div 1.8 = -39.77°C$
   d. 0°F   = __−17.77__ °C $0°F - 32 \div 1.8 = -17.77$

2. Convert the following Celsius temperatures to Fahrenheit.
   a. 25°C   = __77__ °F $(1.8 \times °C) + 32 = 77°F$
   b. 0°C   = __32__ °F $(1.8 \times °C) + 32 = 32°F$
   c. 5°C   = __41__ °F $(1.8 \times °C) + 32 = 41°F$
   d. −10°C = __23.8__ °F $(1.8 \times °C) + 32 = 23.8°F$

Name: _____   Section: _____

PART 2 · CONTINENTALITY AND AIR TEMPERATURE

1. Plot the cities in Table 3.1 on the North America base map provided (Figure 3.2) and label them.

TABLE 3.1 — MONTHLY TEMPERATURE DATA (°C) FOR THREE MIDLATITUDE U.S. CITIES

	J	F	M	A	M	J	J	A	S	O	N	D
San Francisco, California—37.6°N, 122.4°W (San Francisco International Airport)												
Mean temp.	9.2	11.2	11.8	13.1	14.5	16.3	17.0	17.6	18.0	16.1	12.6	9.6
Max. temp.	13.1	15.2	16.0	17.7	19.1	21.2	22.0	22.3	23.1	21.1	16.8	13.3
Min. temp.	5.4	7.2	7.6	8.4	9.8	11.4	12.1	12.7	12.8	11.0	8.3	5.9
Wichita, Kansas—37.7°N, 97.4°W (Wichita/Midcontinent Airport)												
Mean temp.	−0.9	0.8	6.9	13.1	18.3	23.6	26.7	26.1	21.3	14.6	6.8	0.9
Max. temp.	4.5	7.9	13.3	19.9	24.8	30.6	33.6	32.7	27.7	21.5	12.7	6.6
Min. temp.	−6.6	−4.0	0.6	7.0	12.6	18.1	20.8	19.9	15.1	8.3	0.9	−4.2
Norfolk, Virginia—36.9°N, 76.2°W (Norfolk International Airport)												
Mean temp.	3.9	5.0	9.2	13.8	18.9	23.3	25.6	25.1	22.1	16.2	11.3	6.5
Max. temp.	8.5	9.8	14.3	19.3	24.0	28.2	30.2	29.5	26.4	20.8	16.2	11.2
Min. temp.	−0.6	0.1	4.0	8.3	13.7	18.4	21.1	20.7	17.8	11.6	6.5	1.8

(handwritten notes: 167; 158.2; 33.6 − (−6.6); 80.9; 30.2 − (−0.6) = 30.2 + 0.6)

2. Compute the statistics listed below using the data in Table 3.1 and equations 3.3 and 3.5.

	San Francisco, California	Wichita, Kansas	Norfolk, Virginia
Average annual temperature	13.91 °C	13.18 °C	15.07 °C
Maximum range in annual temperature	17.7 °C	40.2 °C	30.8 °C
Month of highest mean temperature	September	July	July

(handwritten: add all the mean temperatures and divided by 12 = ; take the highest temperature and deduct the lowest temperature =)

23

3. Plot the mean monthly temperatures for all three locations on the graph paper provided. Plot temperature on the Y-axis (vertical axis) according to the range of values in the temperature data. Plot months of the year on the X-axis (horizontal axis). Connect the points for each city as a smooth line, and label each line.

4. Based on the statistics computed in question 2, describe the relationship between location and maximum range in temperature. The 3 location in the maps run throughout the west, the middle, and the east coast of the united States. location on the Westcoast is cooler, compared to the East coast

5. Why does San Francisco have a smaller temperature range than Norfolk, Virginia, even though both are located on coasts? Keep in mind the prevailing winds are from the west. because San Francisco have higher temperatures due to to the influence of cold pressure from the Pacific Ocean, coming from the Artic. (NP)

6. Assuming that all three cities receive their highest total monthly insolation in June, how long (in months) is the seasonal lag at:

San Francisco __1 month__ Wichita __3 months__ Norfolk __2 months__

7. a. Which city has the longest seasonal lag in mean monthly temperature? __Kansas__
   b. Why? because of the high temperatures being in the mid west.

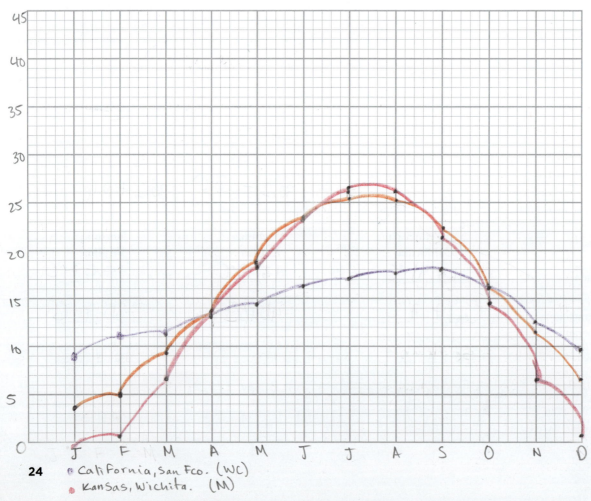

• California, San Fco. (WC)
• Kansas, Wichita. (M)
• Virginia, Norfolk. (EC)

3/19/18 10:53am 64.0°F Sun Angle

FIGURE 3.2 NORTH AMERICA BASE MAP

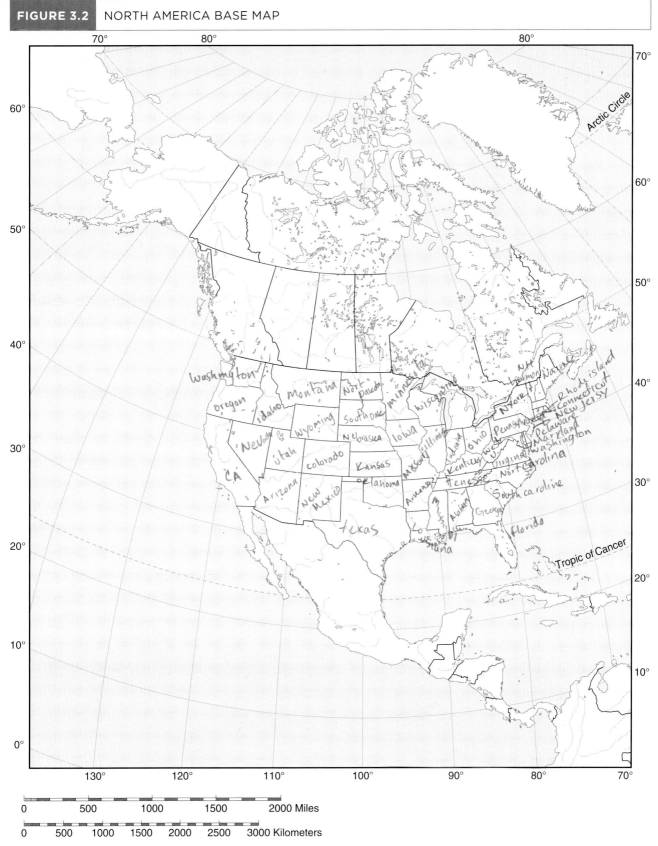

Name: _____   Section: _____

PART 3 · LATITUDINAL TEMPERATURE GRADIENT IN NORTH AMERICA

1. Plot the cities in Table 3.2 on the North America base map provided (Figure 3.2) and label them.

TABLE 3.2	AVERAGE JANUARY AND JULY TEMPERATURES FOR SELECTED CENTRAL NORTH AMERICAN CITIES			
	Latitude	Avg. Temperature (°C)		Temperature
City	(°N)	January	July	Range
Winnipeg, Canada	49.9	−18.6	19.6	
Fargo, North Dakota	46.9	−14.3	21.2	
Sioux Falls, South Dakota	43.6	−10.1	23.5	
Lincoln, Nebraska	40.9	−4.6	25.5	
Manhattan, Kansas	39.2	−2.8	26.2	
Wichita, Kansas	37.7	−1.3	27.4	
Oklahoma City, Oklahoma	35.4	2.1	27.7	
Ft. Worth, Texas	32.8	6.3	29.6	
San Antonio, Texas	29.5	9.6	29.4	
Monterrey, Mexico	25.7	14.7	27.7	

2. a. Plot the January and July average temperatures listed in Table 3.2 against latitude on the graphs provided in Figure 3.3.

   b. Draw a straight line through each temperature graph that best represents the trend of the plotted points. Try to minimize the vertical distance between each point and the trend line.

3. Calculate the January (winter) and July (summer) temperature gradients in degrees Celsius per degree of latitude over the latitudinal range of these cities. Show your work.

   a. January gradient =

   b. July gradient =

27

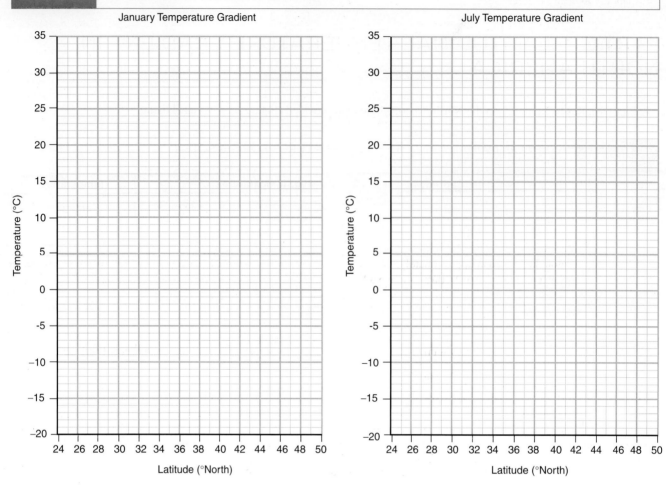

FIGURE 3.3 TEMPERATURE GRADIENT GRAPHS

4. Is the temperature gradient larger or smaller during the winter? _____

5. a. Compute the difference between July average and January average temperatures for each city and enter the value in the space provided in Table 3.2.

   b. Describe the relationship between the temperature range and location between 25°N and 50°N latitude.

6. Hot Springs, Arkansas, has a latitude of about 34°N. Use the graphs you constructed in Figure 3.3 to estimate the average January and July temperatures.

   a. Predicted average January temperature: _____

   b. Predicted average July temperature: _____

   c. Predicted temperature range: _____

   d. Plot Hot Springs on your base map (Figure 3.2).

   e. How do the predicted values compare to the actual values for Hot Springs below?

   Actual average January temperature: 4°C

   Actual average July temperature: 27.6°C

   Actual temperature range: 23.6°C

7. Portland, Oregon, is located at approximately 45°N while Hot Springs, Arkansas, is located at 34°N. Given your answer to (question 5b), would you expect Portland to have a larger or smaller temperature range than Hot Springs?

8. Use the graphs you constructed in Figure 3.3 to estimate the average January and July temperatures at Portland (45°N).

   a. Predicted average January temperature: _____

   b. Predicted average July temperature: _____

   c. Predicted temperature range: _____

   d. Plot Portland on your base map (Figure 3.2).

   e. How do the predicted values compare to the actual values for Portland below?

   Actual average January temperature: 3.9°C

   Actual average July temperature: 19.6°C

   Actual temperature range: 15.7°C

9. a. For which city, Portland or Hot Springs, did your graph best predict the average temperatures and temperature range?

   b. Why? (Hint: Consider where Portland is located and the location of the data used to construct your graphs.)

EXERCISE 4 ✦ ATMOSPHERIC PRESSURE, CIRCULATION, AND WIND

PURPOSE

The purpose of this exercise is to gain a better understanding of the role atmospheric pressure plays in affecting the weather, and to learn how to create and interpret weather maps.

LEARNING OBJECTIVES

By the end of this exercise you should be able to

- create a map and profile of air pressure;
- interpret wind direction and speed from a weather map; and
- explain how pressure gradients and wind velocity change with season.

INTRODUCTION

Air pressure is the force exerted per unit area by the weight of a column of air above a particular point. The length of a column of air above a given reference point decreases as elevation increases, and as a result, air pressure decreases. Because the atmosphere is compressible, density and pressure decrease with height at an increasing rate.

A **barometer** is used to measure air pressure. The mercurial barometer is a very accurate tool for measuring air pressure, and is simply a column of mercury in a glass tube, the mass of the mercury balancing the downward force of the atmosphere. Climatologists and meteorologists use **millibars**, the metric unit for measurement of force. A millibar (mb) is $1/1000^{th}$ of a **bar**—a measurement unit for pressure. One bar is equal to 100,000 pascals, the SI unit for pressure. One millibar is equal to 100 pascals, 0.0295 Hg (mercury height), or 0.0145 psi (pounds per square inch). Mean Sea Level Pressure is equal to 1013.2 mb (29.92 inches of mercury).

Spatial variation of surface air pressure can be induced *thermally* and *mechanically*. Heating and subsequent rising of air draws air molecules away from the surface, decreasing the surface air pressure. Piling up of air molecules, such as when air subsides, can increase air pressure. Regardless of which process is responsible, air pressure changes from place to place.

Mapping and Interpreting Atmospheric Pressure

Information on atmospheric pressure is collected at many weather stations throughout the country. In order to make this information easy to interpret, **isobars,** or lines of equal barometric pressure, are drawn on weather maps. Figure 4.1 provides an example of an atmospheric pressure map with isobars. Notice that the map shows a well-developed low pressure system along the east coast of the United States. An elongate area of high pressure, a ridge, is located over the southwestern United States. Refer to Appendix B to obtain more information about drawing isolines. A **pressure profile** provides a useful aid for visualizing the distribution of air pressure across a map. Figure 4.2 provides an example of a pressure profile for the pressure patterns shown on the map in Figure 4.1. Refer to Appendix C to learn how to draw a profile. The profile axes should always be labeled to indicate what the axes represent, and the units of measurement involved.

Pressure Gradient, Wind Speed, and Wind Direction

Wind is the result of pressure differences in the atmosphere and represents the movement of air molecules to restore equilibrium in the system. The difference in pressure over a unit of distance is known as the **pressure gradient:**

$$\text{Pressure gradient} = \frac{\Delta \text{ Pressure}}{\text{Distance}} \quad (4.1)$$

where Δ (delta) stands for "change in."

Air moves from areas of high pressure to areas of low pressure and thus the pressure gradient affects the direction the wind blows. The steepness of the pressure gradient largely determines the wind speed and can be inferred from the spacing of isobars. Closely spaced isobars indicate a steep pressure gradient, or a large change in pressure over a unit of distance, and strong winds. Widely spaced isobars indicate a gentle pressure gradient and gentle winds (Figure 4.3).

On weather maps, weather stations are represented as a circle. Wind direction is shown by drawing a line in the direction the wind is blowing, and the wind is blowing along the line toward the weather station. Winds are named for the direction they come from, not the direction they are blowing toward. Figure 4.4 shows an example of a west wind.

FIGURE 4.1 UNITED STATES WEATHER MAP WITH ISOBARS

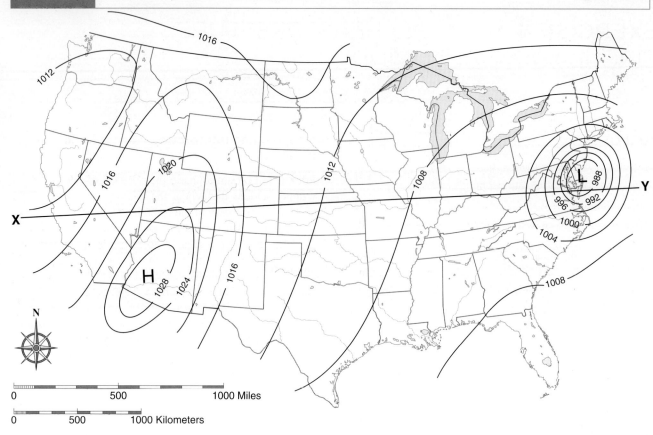

FIGURE 4.2 PROFILE OF ATMOSPHERIC PRESSURE FROM FIGURE 4.1

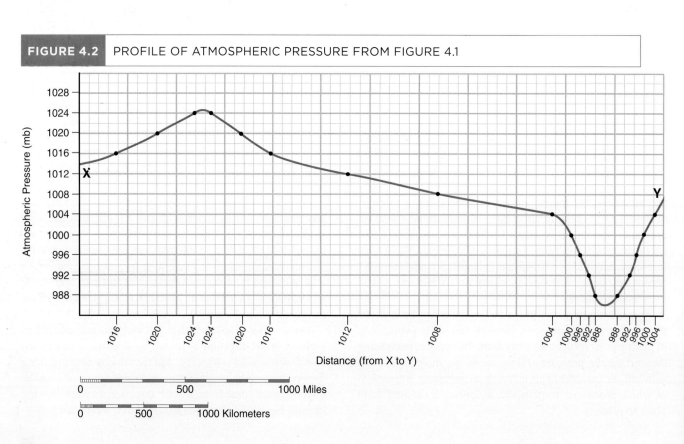

| FIGURE 4.3 | PRESSURE GRADIENTS |

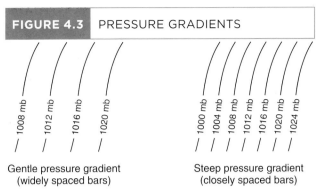

Gentle pressure gradient (widely spaced bars)

Steep pressure gradient (closely spaced bars)

| FIGURE 4.4 | WEATHER STATION WITH A WEST WIND |

Weather maps also show wind speed. An **anemometer** measures wind speed. Wind speed can be reported in either knots or miles per hour (1 knot is equal to 1.152 miles per hour). A series of feathers at the end of the wind direction line indicates the wind speed. The more feathers, the faster the wind speed. Table 4.1 shows the symbols used for indicating wind speed. Note that each symbol represents a range of values. To simplify the symbols for wind speed, use the following generalization:

a half feather equals approximately 5 knots,

a full feather equals approximately 10 knots, and

a triangle equals approximately 50 knots.

IMPORTANT TERMS, PHRASES, AND CONCEPTS

air pressure   isobars
barometer   pressure profile
millibars   pressure gradient
bar   anemometer

| TABLE 4.1 | WIND SPEED SYMBOLIZATION |

Symbol	Statute miles per hour	Knots	Kilometers per hour	Symbol	Statute miles per hour	Knots	Kilometers per hour
◎	Calm	Calm	Calm		44–49	38–42	70–79
	1–2	1–2	1–3		50–54	43–47	80–87
	3–8	3–7	4–13		55–60	48–52	88–96
	9–14	8–12	14–19		61–66	53–57	97–106
	15–20	13–17	20–32		67–71	58–62	107–114
	21–25	18–22	33–40		72–77	63–67	115–124
	26–31	23–27	41–50		78–83	68–72	125–134
	32–37	28–32	51–60		84–89	73–77	135–143
	38–43	33–37	61–69		119–123	103–107	144–196

Name: _____   Section: _____

Isobar: line on a pressure map connecting points of equal elevations/atmospheric pressure

PART 1 † MAPPING AIR PRESSURE

Read Appendix B and Appendix C before attempting Part 1.

1. Draw isobars on the map in Figure 4.5 every 4 millibars, starting with a low of 1000 mb and ending with a high of 1036 mb. Be sure to label each isobar so that the pattern is clear and easy to see.

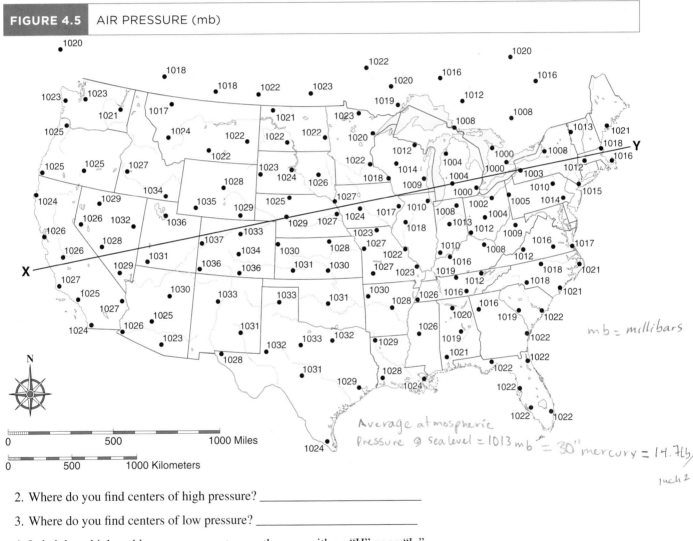

FIGURE 4.5 AIR PRESSURE (mb)

mb = millibars

Average atmospheric pressure @ sea level = 1013 mb = 30" mercury = 14.7 lb/inch²

2. Where do you find centers of high pressure? _____

3. Where do you find centers of low pressure? _____

4. Label these high and low pressure centers on the map with an "H" or an "L."

5. On the graph paper, draw a profile of the pressure gradient from point X, on the west coast, to point Y, on the east coast. Be sure to label the axes properly.

6. In general, wind tends to blow from areas of _____ pressure to areas of _____ pressure.

7. Although differences in atmospheric pressure are the driving force causing winds to blow, winds rarely flow directly down the pressure gradient, because the earth rotates. As a result of the Coriolis force, winds appear to be deflected. Draw several arrows indicating wind direction in and around the high and low pressure centers on your map in Figure 4.5.

8. What were the wind speeds and directions at the following stations on January 3, 1988 (Figure 4.6)?

35

FIGURE 4.6 JANUARY 3, 1988 WEATHER MAP

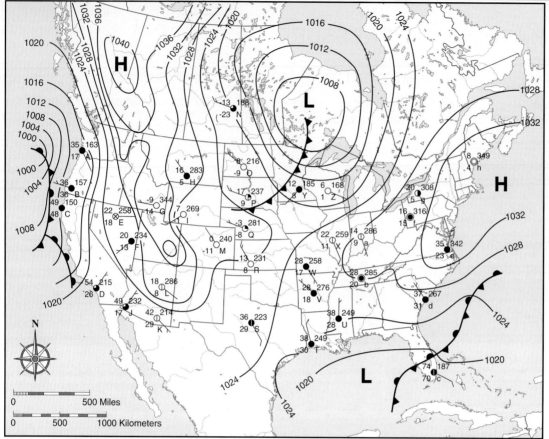

A. Astoria, OR
B. Medford, OR
C. Eureka, CA
D. Los Angeles, CA
E. Winnemucca, NV
F. Ely, NV
G. Pocatello, ID
H. Billings, MT
I. Rock Springs, WY
J. Yuma, AZ
K. Tucson, AZ
L. Winslow, AZ
M. Denver, CO
N. Hudson Bay, Can
O. Bismarck, ND
P. Pierre, SD
Q. North Platte, NE
R. Dodge City, KS
S. Abilene, TX
T. Lake Charles, LA
U. Jackson, MS
V. Little Rock, AR
W. Springfield, MO
X. Springfield, IL
Y. Minneapolis, MN
Z. Wausau, WI
a. Indianapolis, IN
b. Nashville, TN
c. Miami, FL
d. Charleston, SC
e. Norfork, VA
f. Pittsburgh, PA
g. Buffalo, NY
h. Portland, ME

Source: Data from National Weather Service

TABLE 4.2 WIND SPEED AND WIND DIRECTION

Station	Wind Direction	Wind Speed
Dodge City, Kansas		
Wausau, Wisconsin		
Abilene, Texas		
Bismarck, North Dakota		

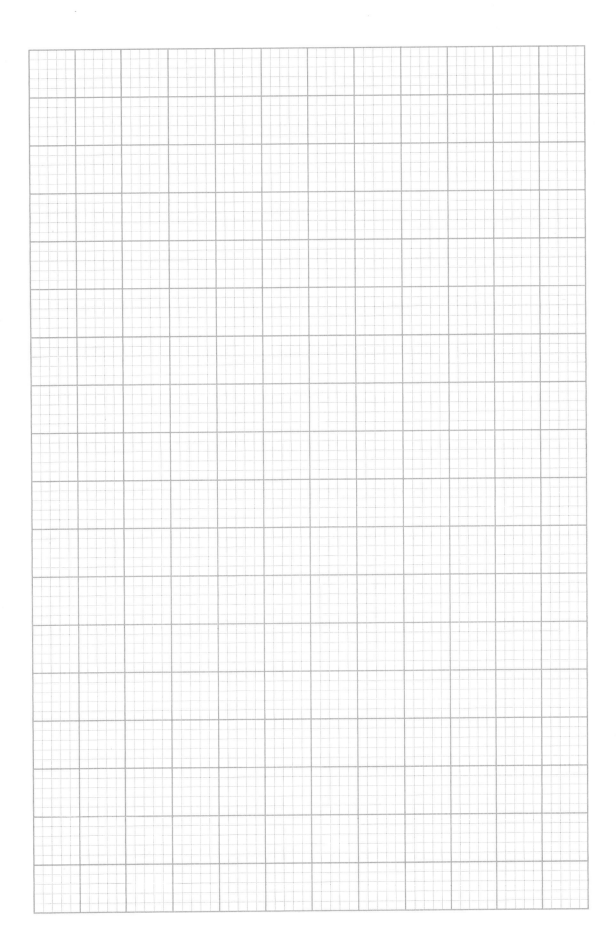

Name: _____ Section: _____

PART 2 † SEASONAL PRESSURE GRADIENTS AND WIND SPEED

Use the January 3, 1988 (Figure 4.6), and July 3, 1991 (Figure 4.7), maps to answer the following questions.

FIGURE 4.7 JULY 3, 1991 WEATHER MAP

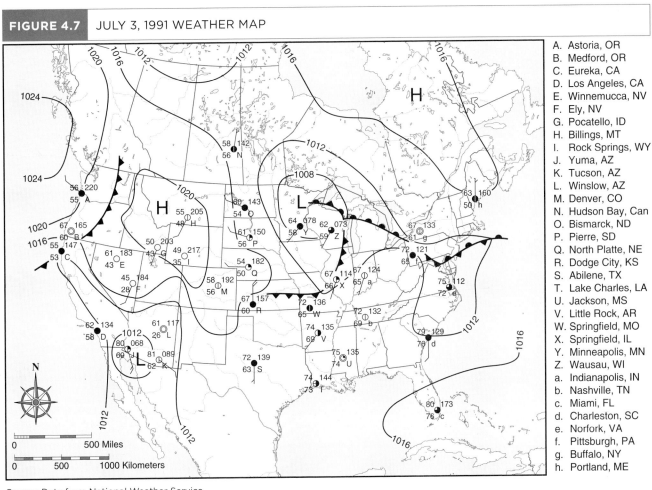

A. Astoria, OR
B. Medford, OR
C. Eureka, CA
D. Los Angeles, CA
E. Winnemucca, NV
F. Ely, NV
G. Pocatello, ID
H. Billings, MT
I. Rock Springs, WY
J. Yuma, AZ
K. Tucson, AZ
L. Winslow, AZ
M. Denver, CO
N. Hudson Bay, Can
O. Bismarck, ND
P. Pierre, SD
Q. North Platte, NE
R. Dodge City, KS
S. Abilene, TX
T. Lake Charles, LA
U. Jackson, MS
V. Little Rock, AR
W. Springfield, MO
X. Springfield, IL
Y. Minneapolis, MN
Z. Wausau, WI
a. Indianapolis, IN
b. Nashville, TN
c. Miami, FL
d. Charleston, SC
e. Norfork, VA
f. Pittsburgh, PA
g. Buffalo, NY
h. Portland, ME

Source: Data from National Weather Service

1. Record the wind speed, in knots, for each of the four stations below using the January 3, 1988 weather map (Figure 4.6). Use the following generalization:

   no feather = 0 knots   full feather = 10 knots

   half a feather = 5 knots   triangle = 50 knots

   Station O _____   Station Y _____

   Station P _____   Station Z _____

2. Record the wind speed for the same four stations as they appear on the July 3, 1991 map (Figure 4.7).

   Station O _____   Station Y _____

   Station P _____   Station Z _____

3. Determine the average wind speed for each map, using only the four stations in questions 1 and 2.

   a. January 3, 1988, average wind speed =

   b. July 3, 1991, average wind speed =

4. Examine these two weather maps. Which month experienced steeper pressure gradients?

5. What evidence from the maps did you use to support your answer to question 4?

6. a. Do the average wind speeds for each map help substantiate your answer to question 4?

   b. How?

7. a. Which season probably had the steeper latitudinal temperature gradient?

   b. Why?

EXERCISE 5 · WATER IN THE ATMOSPHERE

PURPOSE

The purpose of this exercise is to learn about different measurements of humidity and to examine the relationship between humidity and air temperature. The processes of evaporation and condensation are also explored since they affect and are affected by humidity and air temperature.

LEARNING OBJECTIVES

By the end of this exercise you should be able to

- explain the controls over evaporation and humidity;
- explain how humidity is affected by changes in temperature and moisture content; and
- calculate relative humidity.

INTRODUCTION

Humidity, or the amount of water vapor present in the air, plays a role in the heat energy balance of the atmosphere and is important in transporting heat energy around the earth. The amount of water vapor in the air also has an impact on the weather and climatic conditions a location experiences. It is important to understand how to measure humidity since this information is used to assess weather and climate conditions.

Phase Changes of Water

When water undergoes a phase change, the water either absorbs heat energy or releases heat energy. Energy that is absorbed is called **latent heat;** this is heat energy that is stored and cannot be measured or felt. When this energy is released, it is converted to **sensible heat,** or heat which can be felt and measured. Figure 5.1 shows the relative amounts of energy that are absorbed or released as water changes phase.

Evaporation is the phase change from liquid water to water vapor. Evaporation requires the presence of water and absorption of energy from the surrounding environment. If water absorbs heat energy, the surrounding environment will lose it, which is why evaporation is called a "cooling process." For example, soon after you leave the swimming pool on a warm summer day, the water droplets on your skin evaporate. The water droplets absorb energy from the surrounding environment (the air and your skin) to change into a gas. You feel chilled because energy is transferred from your skin to the droplets that evaporate and transfer the stored latent heat into the air. This means that there is a loss of energy from your skin (and air too). It takes about 600 calories of heat to evaporate 1 gram of liquid water.

The rate at which evaporation occurs depends on the availability of water and energy. In general, air temperature affects how much energy is available to turn water from a liquid to a gas. The higher the air temperature, the more energy there is to evaporate water. The amount of water vapor already in the air and the wind speed also affect the evaporation rate. When the maximum concentration of water vapor for a particular temperature is reached, saturation occurs.

Condensation is the phase change from water vapor to liquid water. When water changes from a higher state to a lower state (gas to liquid, liquid to ice, or gas to ice) latent, or stored, energy is released. This is the exact opposite of what happens when water changes from a lower to a higher state.

FIGURE 5.1 ENERGY AND PHASE CHANGES FOR 1 GRAM OF WATER

Latent heat becomes sensible heat. This means the surface on which the water condenses gains the heat energy lost by the water. As a result, on a humid summer day, condensation of water vapor on your skin causes you to feel much warmer than on a dry summer day with the same air temperature.

As with evaporation, air temperature affects the condensation rate. The cooler the air, the less energy there is to maintain water in a gaseous state; hence, condensation occurs more readily as air temperature decreases. In order for water to condense, there generally must be some surface for the water to condense on, and the temperature of this surface also affects the condensation rate. The lower the temperature of a surface, the more likely water will condense on it. In addition, the amount of moisture in the air also affects the rate of condensation.

Humidity, Air Temperature, and Saturation

There are several ways to measure humidity, or the moisture content of the air. Following is a list of the most common measurements of humidity.

- **Specific humidity** (grams/kilogram): the ratio of water vapor present in the air, or the weight of water vapor present, to the total weight of the air; expressed as grams of water vapor per kilogram of moist air.
- **Mixing ratio** (grams/kilogram): the ratio of water vapor present in the air, or the weight of water vapor present, to the total weight of dry air occupying the same space; expressed as grams of water vapor per kilogram of dry air. This is similar to specific humidity, except that specific humidity uses the weight of moist air, while the mixing ratio uses the weight of dry air.
- **Vapor pressure** (millibars): the weight of water vapor in the air; the weight of all the molecules of water vapor in the air excluding the weight of any other molecules of gases composing the air; expressed in millibars.
- **Relative humidity** (percent): the amount of water vapor present in the air relative to the maximum amount the air can hold at a given temperature; expressed as a percent of the maximum amount of water vapor in the air at that temperature.

Temperature directly affects the air's saturation point for water vapor. Water molecules are in a constant state of change between liquid and vapor between the surface and the air. During evaporation, more water molecules change into vapor than back into a liquid (condensation). During condensation, more water molecules are changing into liquid than vapor. Eventually, equilibrium is reached—when evaporation is equal to condensation. At this point, the air is considered saturated. The balance between the two processes can be upset by changing air temperature. Evaporation increases as air temperature increases until a new equilibrium (saturation) is reached with the rate of condensation. At a warmer temperature, the air contains more water at saturation than at a cooler temperature.

Figure 5.2 shows this relationship, using the mixing ratio and air temperature. As air temperature increases, the ability of the air to keep water in a gaseous state also increases. In other words, it takes more water vapor to saturate warm air than it does to saturate cold air. If we know the temperature, we can use Figure 5.2 to determine how many grams of water vapor will saturate the air. For example, if the air temperature is 10°C, move from 10°C on the X-axis vertically up to the saturation curve and then over to the Y-axis for a quantity of 8 grams of water vapor. This is the amount of water vapor required to saturate air at 10°C. Regardless of the measurement of water vapor used, graphs of this sort always show the same pattern: as air temperature

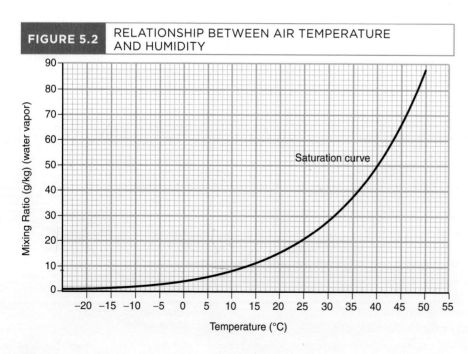

FIGURE 5.2 RELATIONSHIP BETWEEN AIR TEMPERATURE AND HUMIDITY

increases, the ability of the air to keep water in its vapor state also increases.

The **dew point temperature** is the temperature to which air must be cooled in order for it to become saturated. Dew point, therefore, depends on the amount of water vapor present in the air. Thus, if the amount of water vapor present in the air is known, Figure 5.2 can be used to determine the dew point temperature. For example, to determine the dew point temperature of the air with a mixing ratio of 20 g/kg, find 20 g/kg on the Y-axis of the graph, move horizontally to the saturation curve, and then down to the X-axis for a dew point temperature of 25°C.

We can also use Figure 5.2 to calculate the relative humidity, provided we know the current air temperature and the current water vapor content of the air. For example, if the air has a mixing ratio of 10g/kg and the air temperature is 20°C, we can use Figure 5.2 to determine the quantity of water vapor required to saturate the air. In this case, that would be 15g/kg. The relative humidity is the ratio of the water vapor present to the water vapor required to saturate the air at the current temperature, or 10g/kg divided by 15g/kg, which results in a relative humidity of 67%.

One instrument used to measure humidity is a **sling psychrometer.** A sling psychrometer consists of two identical thermometers mounted side-by-side. The **wet-bulb thermometer** consists of a thermometer with a cloth wick saturated with distilled water tied to one end. The **dry-bulb thermometer** is exposed directly to the air. Swinging the sling psychrometer allows a continuous flow of air to pass across the saturated cloth wick of the wet-bulb thermometer. Water evaporates from the wick, causing the temperature of the wet-bulb to decrease. The heat used to evaporate water is lost from the wick surrounding the wet-bulb and this causes the thermometer reading to drop. The amount of cooling that takes place is directly related to the dryness of the air. The drier the air, the more wet-bulb cooling occurs. Therefore, the larger the difference between dry- and wet-bulb temperatures, the lower the relative humidity. If the air is saturated, no cooling takes place, and the two thermometers have the same temperature readings.

To determine the relative humidity and the dew point temperature with a sling psychrometer, use Tables 5.1 and 5.2. First, record the air (dry-bulb) temperature. Then calculate the difference between the dry- and wet-bulb temperatures, or the **wet-bulb depression.** For example, if the dry-bulb temperature is 20°C and the wet-bulb is 15°C, then the wet-bulb depression is 5°C. To determine the relative humidity, find the dry-bulb temperature on the left-hand column of Table 5.2 and the wet-bulb depression along the top. The relative humidity is found where the column and row meet. In this example, the relative humidity is 58%. The dew point temperature is determined in the same way using Table 5.1. In this case it would be 12°C.

IMPORTANT TERMS, PHRASES, AND CONCEPTS

humidity	vapor pressure
latent heat	relative humidity
sensible heat	dew point temperature
evaporation	sling psychrometer
condensation	wet-bulb thermometer
specific humidity	dry-bulb thermometer
mixing ratio	wet-bulb depression

TABLE 5.1 — DEW POINT TEMPERATURE (°C)

Dry-Bulb (Air) Temperature (°C)	Wet-Bulb Depression (Dry-Bulb Temperature − Wet-Bulb Temperature) (°C)															
	0.5	1.0	1.5	2.0	2.5	3.0	3.5	4.0	4.5	5.0	7.5	10.0	12.5	15.0	17.5	20.0
−20.0	−25															
−17.5	−21	−33														
−15.0	−19	−27	−38													
−12.5	−15	−23	−28													
−10.0	−12	−18	−22	−29												
−7.5	−9	−14	−18	−21												
−5.0	−7	−11	−14	−17	−27											
−2.5	−4	−8	−10	−13	−20	−36										
0	−1	−6	−7	−9	−16	−26	−34									
2.5	1	−3	−4	−6	−11	−19	−24									
5.0	4	0	−1	−3	−8	−14	−17	−22	−28							
7.5	6	3	2	0	−4	−10	−12	−15	−19	−24						
10.0	9	6	4	3	−1	−6	−8	−10	−13	−16						
12.5	12	8	7	6	2	−3	−4	−6	−8	−10	−48					
15.0	14	11	10	9	5	1	−1	−2	−4	−6	−22					
17.5	17	13	12	12	8	4	2	1	0	−2	−28					
20.0	19	16	15	14	11	7	6	4	3	2	−7	−28				
22.5	22	18	18	17	13	10	9	8	7	5	−2	−14				
25.0	24	21	20	20	16	12	12	11	10	8	2	−7	−35			
27.5	27	24	23	22	19	15	14	14	13	12	6	−1	−15			
30.0	29	26	26	25	21	18	17	16	16	15	10	3	−6	−38		
32.5	32	29	28	27	24	21	20	19	18	18	13	7	0	−14		
35.0	34	31	31	30	27	23	23	22	21	20	16	11	5	−5	−32	
37.5	37	34	33	32	29	26	25	25	24	23	19	14	9	2	−11	
40.0	39	36	36	35	32	29	28	27	26	26	22	18	13	7	−2	0
		39	38	38	34	31	31	30	29	28	25	21	16	11	4	6
					37	34	33	32	32	31	28	24	20	15	9	
						36	36	35	34	34	30	27	23	18	13	

TABLE 5.2 RELATIVE HUMIDITY (%)

	Wet-Bulb Depression (Dry-Bulb Temperature	Wet-Bulb Temperature) (°C)															
Dry-Bulb (Air) Temperature (°C)	0.5	1.0	1.5	2.0	2.5	3.0	3.5	4.0	4.5	5.0	7.5	10.0	12.5	15.0	17.5	20.0
−20.0	70	41	11													
−17.5	75	51	26	2												
−15.0	79	58	38	18												
−12.5	82	65	47	30	13											
−10.0	85	69	54	39	24	10										
−7.5	87	73	60	48	35	22	10									
−5.0	88	77	66	54	43	32	21	11	1							
−2.5	90	80	70	60	50	42	37	22	12	3						
0	91	82	73	65	56	47	39	31	23	15						
2.5	92	84	76	68	61	53	46	38	31	24	1					
5.0	93	86	78	71	65	58	51	45	38	32	11					
7.5	93	87	80	74	68	62	56	50	44	38	19	4				
10.0	94	88	82	76	71	65	60	54	49	44	25	12	2			
12.5	94	89	84	78	73	68	63	58	53	48	31	18	8			
15.0	95	90	85	80	75	70	66	61	57	52	36	24	14	1		
17.5	95	90	86	81	77	72	68	64	60	55	40	28	19	17	1	
20.0	95	91	87	82	78	74	70	66	62	58	44	32	23	12		
22.5	96	91	87	83	78	74	70	66	62	58	44	32	23	12		
25.0	96	92	88	84	81	77	73	70	66	63	50	39	30	20	11	1
27.5	96	92	89	85	82	78	75	71	68	65	52	42	33	23	14	6
30.0	96	93	89	86	82	79	76	73	70	67	54	44	36	26	18	10
32.5	97	93	90	86	83	80	77	74	71	68	56	46	38	29	21	13
35.0	97	93	90	87	84	81	78	75	72	69	56	46	38	29	21	13
37.5	97	94	91	87	85	82	79	76	73	70	58	48				
40.0	97	94	91	88	85	82	79	77	74	72	59	48	38	29	21	13

Name: _____ **Section:** _____

MOISTURE PHASE CHANGES AND MEASURING HUMIDITY

1. When water freezes, is heat released or absorbed by the ice?

   Heat is released ✓

2. How many calories would it take to evaporate 80 grams of water?

   (600 cal / 1 gram of H_2O)

   80 gm × 600 cal/gm = 48,000 calories ✓

3. When ice melts, would you expect the temperature of the surrounding air to increase or decrease? Why?

   Increase, because the surrounding air will absorbed the water vapor

   decrease → water absorbs heat −.5

4. Many environmental factors influence the rate of evaporation from an open water surface. List three factors that could be important.

   a. Temperature (water/air)

   b. Degree of wind ✓

   c. Humidity

5. If the amount of water vapor in the air stays constant and the temperature of the air decreases, will the relative humidity increase, decrease, or stay the same? Increase the relative humidity.

6. If the amount of water vapor in the air increases and the temperature of the air stays constant, will the vapor pressure increase, decrease, or stay the same?

7. If the amount of water vapor in the air stays constant and the temperature of the air increases, will the specific humidity increase, decrease, or stay the same?

8. If the amount of water vapor in the air stays constant and the temperature of the air decreases, will the saturation mixing ratio increase, decrease, or stay the same?

9. If the amount of water vapor in the air decreases and the temperature of the air stays constant, will the dew point temperature increase, decrease, or stay the same?

10. Using a sling psychrometer, you measure an air temperature of 30°C and a wet-bulb temperature of 25°C.

    a. What is the wet-bulb depression? ___5°C___

    b. What is the dew point temperature? ___23°C___

    c. What is the relative humidity of the air? ___63° 67%___ −.25

a). Dry bulb 30°C
    wet bulb 25°C
    _____
              5°C

−2.25

47

11. Use a sling-psychrometer to calculate the dew point temperature and the relative humidity in the classroom.
    a. What is the air temperature? __21°C__
    b. What is the wet-bulb temperature? __15°C__
    c. What is the wet-bulb depression? __6°C__
    d. What is the dew point temperature? __11°C__
    e. What is the relative humidity of the air? __52°C__

    dry bulb: 21°C
    wet bulb: 15°C
    ―――――
    6°C

Use Figure 5.2 to answer the following questions.

12. If the air temperature is 25°C, how much water vapor does it take to saturate the air?

    20 g/kg ✓

13. If the air temperature is 5°C, how much water vapor does it take to saturate the air?

    6 g/kg ✓

14. If the air temperature increases from 25°C to 35°C, how much more water vapor will it take to saturate the air?

    17–18° g/kg ✓

15. If the air temperature increases from 5°C to 10°C, how much more water vapor will it take to saturate the air?

    8 g/kg   2 g/kg −.5

16. If the water vapor content of the air is 12 g/kg and the air temperature is 30°C, what two specific actions will make the air saturated?
    a. temperature goes down ✓
    b. evaporation

17. If the water vapor content of the air is 8 g/kg and the air temperature is 30°C, what is the relative humidity?

    $\frac{8g}{28g} \times 100 = 28.57\%$

    water vapor: 8 g/kg
    Air temp: 30°C ✓

18. If the relative humidity of the air is 60% and the temperature of the air is 10°C, how much water vapor is present in the air?

    $8 \left(\frac{x}{8 g/kg} \times 100 \right) = \left((60\%) \div 100 \right) \cdot 8$

    $x = 4.8 g$

    water vapor: .
    Air temperature: 10°C
    Relative Humidity: 60%

19. Cold, continental polar air is often described as being dry even when its relative humidity is very high. Why is this so?

    because it cannot hold humidity, due to the cold. ✓

20. A snowstorm is occurring in Chicago, Illinois, and the air temperature is −9°C and the dew point is −9°C. Meanwhile, in the Atacama Desert (in Chile, along the west coast of South America) the air temperature is 35°C and the dew point is 7°C.

    is a 100% humidity

    a. Which location has the highest relative humidity? __Chicago, Illinois__ ✓
    b. Which location has the highest vapor pressure? __Atacama, Chile__ ✓

48

46.5

EXERCISE 16 — INTRODUCTION TO TOPOGRAPHIC MAPS

PURPOSE

The purpose of this exercise is to learn how to use and interpret topographic maps.

LEARNING OBJECTIVES

By the end of this exercise you should be able to

- draw contour lines from spot elevations;
- use a representative fraction to calculate distances separating points on a map;
- calculate elevation gradients from topographic maps;
- interpret contour lines correctly, including determining their elevation using the contour interval, assessing gradient based on contour line spacing, and identifying hills, valleys, and other landforms based on contour line patterns;
- draw a topographic profile with a given vertical exaggeration;
- describe the color coding used for categorizing topographic map symbols; and
- differentiate small-scale maps from large-scale maps and describe how the scale of a map affects the representation of the landscape.

INTRODUCTION

Topographic maps are important tools for geographers, earth scientists, geologists, engineers, planners, and a variety of other scientists. Maps provide a scale representation of the earth's features. Since map features are always drawn smaller than the actual features, cartographers must be selective when deciding what information to include or exclude from a map. The purpose of a map is a primary factor guiding cartographers when making these decisions. The purpose of topographic maps is to depict the physical features of the earth (e.g., topography, elevation, rivers, and wetlands). Cultural features such as county boundaries, roads, and buildings are also included because they serve as important reference points.

Map Scale

Since features on maps are shown at a reduced size, it is necessary to state how the information on a map relates to the real world. **Map scale** represents the ratio between the distance covered on a map and the actual distance covered on the earth.

There are three ways of expressing map scale. First, map scale may be expressed as a **statement,** such as "one inch equals one mile." Second, map scale may be presented as a **graphic** scale, such as that shown in Figure 16.1. Third, map scale may be expressed as a **representative fraction** or **ratio,** such as 1:24,000. This representative fraction means that one unit of measurement on the map is equal to 24,000 units of the same measurement on the earth's surface. These scales are set up for linear measurements. Topographic maps contain both a representative fraction and a graphic scale.

Any unit of measurement may be used with a representative fraction. For example, on a map with a 1:24,000 scale:

1 inch on the map = 24,000 inches on the earth, and

1 centimeter on the map = 24,000 centimeters on the earth.

The representative fraction is versatile; thus, it is important to know how to use this type of scale. Suppose we measure a distance of four inches between two points on a map with a 1:24,000 scale. If we want to convert this map distance of four inches to an actual distance on the earth's surface in miles, four conversion factors are necessary (for more information on conversion factors, see Appendix A):

1 map inch = 24,000 earth inches (the map scale),

12 inches = 1 foot,

5280 feet = 1 mile, and

63,360 inches = 1 mile.

There are two methods for converting this linear measurement in inches on a map to miles on the earth's surface:

(1) Convert the map distance to an earth distance using the representative fraction:

$$X \text{ mi} = \frac{4 \text{ in}_{map}}{1} \times \frac{24{,}000 \text{ in}}{1 \text{ in}_{map}} \times \frac{1 \text{ ft}}{12 \text{ in}} \times \frac{1 \text{ mi}}{5280 \text{ ft}} = 1.5 \text{ mi}$$

(16.1)

FIGURE 16.1 GRAPHIC MAP SCALE

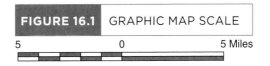

(2) Change the units associated with the earth distance of the representative fraction from inches to miles:

$$X \text{ mi} = \frac{24{,}000 \text{ in}}{1 \text{ in}_{map}} \times \frac{1 \text{ ft}}{12 \text{ in}} \times \frac{1 \text{ mi}}{5280 \text{ ft}} = 0.38 \text{ mi} \quad (16.2)$$

The map scale can now be rewritten as: $1 \text{ in}_{map} = 0.38 \text{ mi}_{earth}$.

Then convert the map distance to an earth distance in miles using the new version of the map scale:

$$X \text{ mi} = \frac{4 \text{ in}_{map}}{1} \times \frac{0.38 \text{ mi}}{1 \text{ in}_{map}} = 1.5 \text{ mi} \quad (16.3)$$

To compare information from two maps accurately, it is necessary to know the scale of each map. If two maps are drawn on the same size piece of paper and they have different scales, the smaller scale map refers to the map, which covers a larger area of the earth's surface. As a result, features on this map look smaller than the same features on a larger scale map (Figure 16.2). Everything must be drawn smaller in order to fit all the information onto the piece of paper. The larger scale map does not cover as much ground area, and features may be drawn larger. Map scale also affects the amount of detail included on a map. A larger scale map may include more specific, detailed information on a particular region than a smaller scale map of the same region. In summary:

- on **small-scale maps** features look small, the ground area covered is large, and few details are shown;
- on **large-scale maps** features look large, the ground area covered is small, and many details are shown.

Topography

One of the primary purposes of topographic maps is to show topography — the shape of the landscape as revealed by changes in elevation. This is accomplished by drawing **contour lines,** lines of equal elevation. Contour lines are drawn as solid brown lines. As with other types of isolines, contour lines may not cross. **Index** contour lines are thicker than **intermediate** contour lines and are labeled with elevations. The **contour interval** is the vertical distance separating successive intermediate contour lines and is stated in the bottom margin of the map. In very flat regions, **supplementary** contour lines may be included to provide more accurate information on the relief. Supplementary contours are generally half the contour interval and are dotted brown lines. **Depression** contours are hachured, and **bathymetric** contours, which show the depth of water bodies, are drawn in blue and do not follow the contour interval.

To determine the elevation of a point located on an index contour, find the elevation label associated with that contour line. For points located on intermediate contours, find the nearest index contour and use the contour interval to count up or down in elevation. The elevation of points between contour lines must be estimated. Spot elevations of survey markers called **benchmarks** are included on topographic maps. Benchmarks are designated with a "BM," an "x," or a triangle, and are permanent markers placed by the U.S. Geological Survey. These elevations are shown in black type, as opposed to brown type for contour line elevations, and are used to help draw contours but are not necessarily associated with a specific contour line.

FIGURE 16.2 TOPOGRAPHIC MAPS OF ARPIN, WISCONSIN

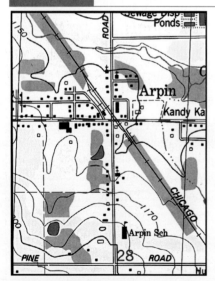
(a) Scale 1:24,000
Area = 0.57 mi²
Much detail

(b) Scale 1:62,500
Area = 3.89 mi²

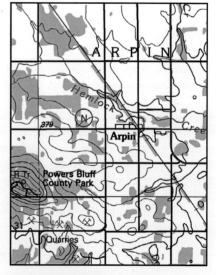
(c) Scale 1:100,000
Area = 9.96 mi²
Little detail

Source: Maps courtesy of the USGS.

Contour lines depict the relief of an area. **Local relief** refers to the difference in elevation between adjacent hills and valleys. **Total relief** is the difference in elevation between the highest and lowest points on a map. Determining elevations and differences in elevation between points is only part of assessing the topography of a region.

Interpreting contour lines and the "picture" they show takes practice. Closed contour lines represent hills, mountains, or high points (Figure 16.3 a and c). A good estimate of the elevation on a hill top or ridge top is to add half the contour interval to the elevation of the inner-most closed contour line representing the hill or ridge. Depressions are represented as closed contours with hachure marks drawn toward the center of the depression (Figure 16.3 b). To estimate the bottom of a depression, subtract half the contour interval from the elevation of the inner-most closed, hachured contour line. Contour lines of the same elevation are repeated on either side of ridges and valleys. If we walk down into a valley and then walk up the other side, we cross the same elevation twice—once on the way down and once on the way up. The same is true of ridges (Figure 16.3 c and d).

There are two other important characteristics of contour lines. First, contour lines crossing rivers or valleys always bend upstream, or upvalley, creating "V's" (Figure 16.4). These V's always point upstream (toward higher elevation).

Second, the spacing of contour lines indicates the gradient of a slope. **Gradient** is the change in elevation per unit distance:

$$\text{Gradient} = \frac{\Delta \text{ Elevation}}{\text{Distance}} \text{ where } \Delta \text{ (delta) stands for "change in."} \quad (16.4)$$

If we want to calculate the gradient between the two benchmarks shown in Figure 16.3 (a) there are three steps to follow:

(1) Determine the change in elevation between the two points: In Figure 16.3 (a), the two benchmarks have elevations of 500 feet and 460 feet:

$$500 - 460 = 40 \text{ ft} = \Delta \text{ elevation} \quad (16.5)$$

(2) Calculate the distance separating these two points, using the conversion procedure outlined in equation 16.1 or equations 16.2 and 16.3:

The map distance between the two benchmarks in Figure 16.3 (a) is 0.375 inches. If this map has a scale of 1:62,500, the earth distance separating these two points is:

$$X \text{ mi} = \frac{0.375 \text{ in}_{\text{map}}}{1} \times \frac{62,500 \text{ in}}{1 \text{ in}_{\text{map}}} \times \frac{1 \text{ ft}}{12 \text{ in}} \times \frac{1 \text{ mi}}{5280 \text{ ft}} = 0.37 \text{ mi}$$

(16.6)

FIGURE 16.3 CONTOUR LINES SHOWING DIFFERENT LANDFORMS

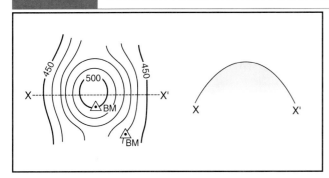

(a) Hill

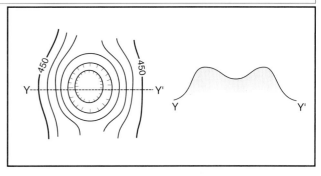

(b) Depression

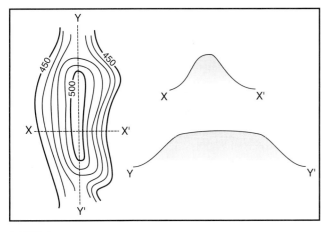

(c) Ridge

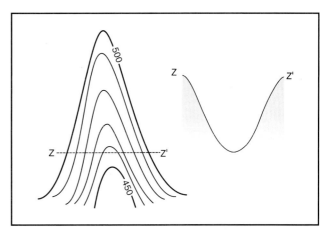

(d) Valley

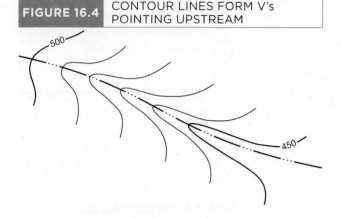

FIGURE 16.4 CONTOUR LINES FORM V's POINTING UPSTREAM

(3) Calculate the gradient by dividing the change in elevation by the distance (equation 16.4):

$$\text{Gradient} = \frac{\Delta \text{Elevation}}{\text{Distance}} = \frac{40 \text{ ft}}{0.37 \text{ mi}} = 108 \text{ ft/mi} \quad (16.7)$$

Although gradient may be calculated using this formula, contour lines allow comparison of the *relative* gradient at different places on a map by their spacing: the closer the contour lines, the steeper the gradient (slope), and the farther apart the contour lines, the gentler the gradient (slope).

Often geographers want a picture of the shape of the landscape along a selected line and they create this by drawing a **topographic profile.** This profile helps visualize the relief of an area. It shows what we would see if we walked along that line. Appendix C outlines the steps for drawing a profile from an isoline map. In addition to these steps, when drawing a topographic profile it is necessary to decide how much to exaggerate the vertical relief. If the horizontal and vertical axes are drawn to the same scale, most topographic profiles will show a flat landscape, which defeats the purpose of drawing a profile—to help us visualize the landscape. The amount of **vertical exaggeration** determines how "stretched out" the relief appears on the profile. Figure 16.5 provides an example of a profile drawn with two different vertical exaggerations. The scale on the horizontal axis is the same as the scale on the map. A vertical exaggeration of 10X means that the distance on the vertical axis representing one mile will be 10 times greater than the distance representing one mile on the horizontal axis; for example, if 1 inch = 1 mile on the horizontal axis, then 10 inches = 1 mile on the vertical axis. The steps for determining vertical exaggeration are:

(1) Determine the map scale. For example, if the map scale is:

1:24,000   1 map in = 24,000 earth in.   (16.8)

(2) Convert the units on the right side of the representative fraction to either feet or meters, depending on the units representing elevation on the map. If our map shows elevations in feet, we will convert the 24,000 earth inches to earth feet:

$$24{,}000/12 = 2000 \text{ ft.} \quad (16.9)$$

We can now state the scale as:

1 map in = 2000 earth ft.   (16.10)

And as a result:

1 horizontal graph in = 2000 earth ft.   (16.11)

This lets us know that the scale of our map is also the scale for the horizontal axis of our profile. For every inch we travel across our profile, the actual distance covered is 2000 feet.

(3) To determine the vertical exaggeration (the scale for the vertical axis), replace the 1 on the left side of equation 16.11 with whatever the vertical exaggeration is. For example:

for a 10X vertical exaggeration, 10 vertical graph in = 2000 earth ft;   (16.12)

for a 20X vertical exaggeration, 20 vertical graph in = 2000 earth ft.   (16.13)

(4) Simplify the vertical exaggeration so that you know what 1 inch on the vertical axis equals:

10X V.E.: $\dfrac{10 \text{ vertical graph in}}{10}$

$= \dfrac{2000 \text{ earth ft.}}{10}$ (16.14)

thus 1 vertical graph in = 200 earth ft   (16.15)

20X V.E.: $\dfrac{20 \text{ vertical graph in}}{20}$

$= \dfrac{2000 \text{ earth ft.}}{20}$ (16.16)

thus 1 vertical graph in = 100 earth ft.   (16.17)

To label the vertical axis, start with the minimum elevation along the profile, or some convenient number less than that. For a 10 times vertical exaggeration, increase the elevation by 200 feet for every inch, or for a 20 times vertical exaggeration by 100 feet for every inch, until the highest elevation, or some convenient elevation higher than that, is reached.

Once the vertical exaggeration is selected and the vertical axis is properly labeled, continue drawing the profile according to the directions in Appendix C. The map scale must be included on the profile, and the vertical exaggeration must be stated. Other information may be included, such as the location of rivers, lakes, or cultural features.

The profile in Figure 16.5 shows the highest point, and examination of the topographic map reveals this is where there are some closed contour lines. The profile also shows changes in gradient. The steepest gradients are associated with the closest contour lines, while those areas on the map with widely spaced contour lines show up on the profile as being less steep.

FIGURE 16.5 CONSTRUCTION OF A TOPOGRAPHIC PROFILE WITH DIFFERENT VERTICAL EXAGGERATIONS

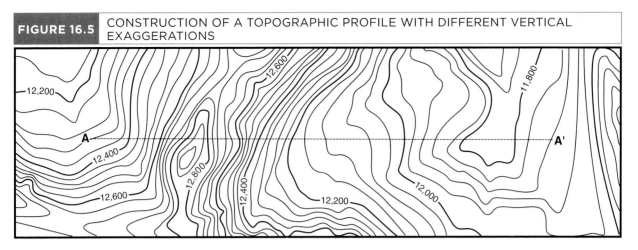

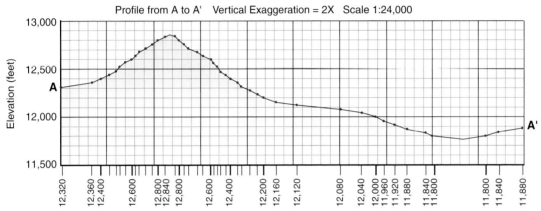

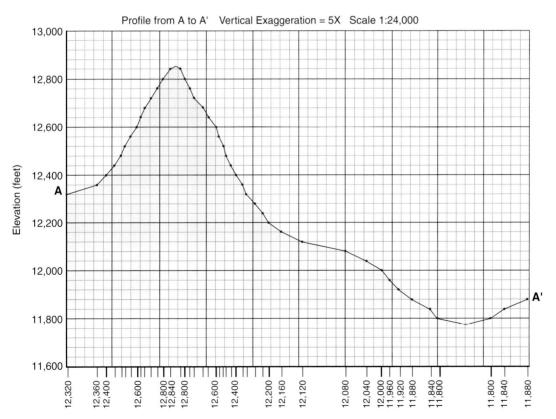

Topographic Map Symbols

Topographic maps show a variety of other physical landscape features and human landscape features in addition to topography through a standard set of map symbols. Appendix E, Figure E.25 contains the U.S. Geological Survey's (USGS) key to topographic map symbols. Since there are a lot of symbols, it is helpful to remember some basics regarding the use of color on topographic maps. Green shading represents forested areas. Red shading represents urban areas. Blue shading and blue lines represent water. Black lines represent political boundaries or roads. Red lines represent survey lines. Brown lines are contour lines. Purple features represent updated information. Construction of colored maps requires a separate map for each color. These separate maps are then combined to create the final map. Maps are periodically updated to show changes in land use, roads, or natural features (such as rivers). Since updating the separate maps for each color and then recombining them is expensive, the USGS creates a new map with the updates and prints this updated map in purple along with the other map layers. Thus, any type of feature may appear in purple on an updated map.

IMPORTANT TERMS, PHRASES, AND CONCEPTS

map scale: statement, graphic, ratio

representative fraction

small scale/large scale maps

contour interval

benchmark

local relief

contour lines: index, intermediate, supplementary, depression, bathymetric

total relief

gradient

topographic profile

vertical exaggeration

PART 1 | DRAWING CONTOUR LINES

Draw contour lines for the spot elevations shown, using a 10-foot contour interval. Label every contour line. Refer to Appendix B for information on drawing isolines. The 950-foot contour line is drawn for you. The lines shown with a dash and three dots represent rivers.

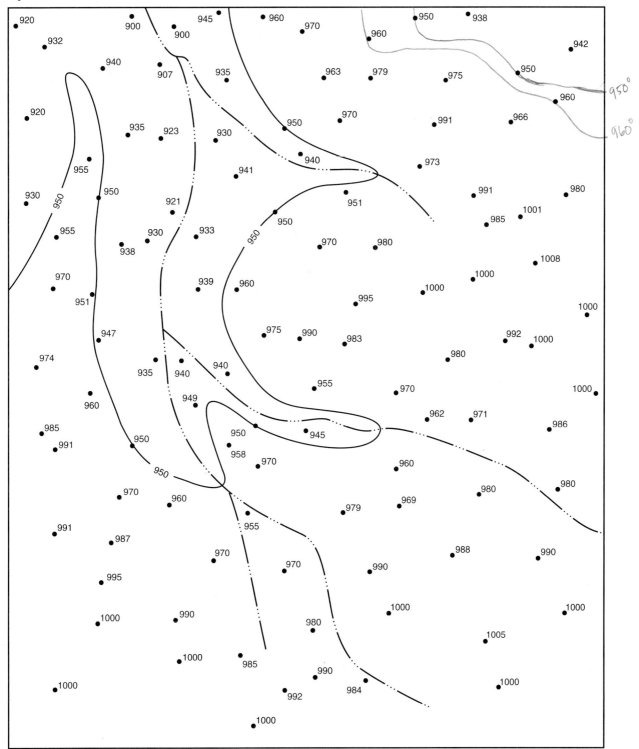

Name: _____   Section: _____

PART 2 • TOPOGRAPHIC MAP INTERPRETATION

Use the topographic maps of Thousand Springs, Idaho (Appendix E, Figure E.7), and Hartford, Alabama (Appendix E, Figure E.8), and the key to topographic map symbols (Appendix E, Figure E.25) to answer these questions.

1. What is the scale for each of these maps?

   Thousand Springs, Idaho   __1:24,000__

   Hartford, Alabama   __1:24,000__

2. What is the contour interval for each of these maps?

   Thousand Springs, Idaho   __20 ft__

   Hartford, Alabama   __10 ft__

3. What is the total relief (difference between highest and lowest elevations) for each of these maps?

	Thousand Springs, Idaho	Hartford, Alabama
a. Highest elevation on the map:	3440 ft	285 ft
b. Lowest elevation on the map:	2880 ft	155 ft
c. Total relief:	6,320 ft	440 ft

4. The two maps have the same scale but different contour intervals.

   a. Why are the contour intervals different? (Hint: Consider the total relief.)

   Idaho, has a lot more steeper contours due to the highest elevation compared to Alabama's lowest elevation.

   b. What would the Thousand Springs, Idaho, map look like if it had the same contour interval as the Hartford, Alabama, map?

   The contour lines on the Idaho map would have a lot more steep elevation and will be harder to read, because the contour lines will be very close together.

5. Calculate the following three gradients in feet per mile, using equation 16.4. Round your answers off to one decimal place. Show all your work!

   a. from point A to point B on the Hartford, Alabama, map

   A = 280' ft   (Distance on map from A to B = 1.25")
   B = 250' ft   1.25" × 24,000 = 30,000 ÷ 12" ÷ 5280 ft
   Change in elevation is 30 ft   = 0.47 mile

   Gradient = 30 ft / 0.47 mi = 63.83 ft/mile

   0.47 mile

   b. from point A to point B on the Thousand Springs, Idaho, map
   A = 2880 ft   Distance on map A to B: 1587.3 ft/m
   B = 3180 ft   0.5" × 24,000 = 12,000 ÷ 12" ÷ 5280 ft = 0.189 miles
   Change in elevation is: 300 ft

   300 ft / 0.189 m = 1587.3 ft/mile

   c. from point E to point F on the Thousand Springs, Idaho, map
   Distance on map E to F: 119.2 ft/m
   E: 3420 ft   4.875 × 24,000 = 117,000 In ÷ 12 = 9,750 ft ÷ 5280 ft/mile = 1.846
   F: 3,200 ft   Inches
   Change in elevation is: 220 ft

   220 ft / 1.846 m = 119.2 ft/miles

6. Compare the three gradients you calculated in question 5 with the spacing of the contour lines associated with each location. Classify the contour line spacing as wide, moderate, or close.

	Gradient	Contour Line Spacing
a. Hartford (A to B)	6383 ft/mile	10 ft
b. Thousand Springs (A to B)	1587.3 ft/mile	20 ft
c. Thousand Springs (E to F)	119.2 ft/mile	20 ft

d. What relationship exists between gradient and contour line spacing?

the higher the gradient the closest together the contour lines are.

7. Draw a topographic profile along the line drawn from point C to point D on the Thousand Springs, Idaho, map (Appendix E, Figure E.7). Use just the index contour lines to draw this profile. Use the graph paper in Figure 16.6. Refer to Appendix C for information on drawing profiles. Determine the range of elevations the vertical axis needs to cover and label the axis so that the vertical exaggeration is 20X. The map scale and the scale of the horizontal axis on the graph is 1:24,000 or, 1 inch on the map or graph equals 2000 feet on the earth. According to equations 16.6 and 16.7, this means that 1 inch on the vertical axis equals 100 feet. The lowest elevation is marked on the graph (Figure 16.6). Finish labeling the vertical axis, increasing elevation 100 feet every time you move up 1 inch. Mark the location of the Snake River on your profile.

8. Draw a topographic profile along the line drawn from point C to point D on the Hartford, Alabama, map (Appendix E, Figure E.8). For this profile, use every contour line, not just the index contour lines. Use the graph paper in Figure 16.7. Set up the vertical axis with a 20X vertical exaggeration the same way you did for question 7 and label the vertical axis accordingly. Mark the location of Hurricane Creek on your profile.

9. Figures 16.8 and 16.9 have other versions of the topographic profile for Thousand Springs, Idaho. These profiles were drawn along the same line as the profile you drew for question 7, but Figure 16.8 has a 10X vertical exaggeration, while Figure 16.9 has only a 2X vertical exaggeration. Compare Figures 16.8 and 16.9 with your profile from Thousand Springs, Idaho (Figure 16.6).

   a. Does the change in vertical exaggeration affect our ability to determine where hills, valleys, and other topographic features are located?

   b. Does the change in vertical exaggeration affect how steep the landscape looks on the profile? (Think if you were planning to hike along this line—does one profile suggest an easier hike than the other?)

   c. What do you think your profile from Hartford, Alabama, would look like if you had drawn it with only a 2X vertical exaggeration?

   d. If you want to compare the landscape in Hartford, Alabama, with the landscape in Thousand Springs, Idaho, which Thousand Springs profile would be more appropriate, the one you drew (Figure 16.6), Figure 16.8, or Figure 16.9? Why?

FIGURE 16.6 TOPOGRAPHIC PROFILE FROM C TO D, THOUSAND SPRINGS, IDAHO. VERTICAL EXAGGERATION IS 20X

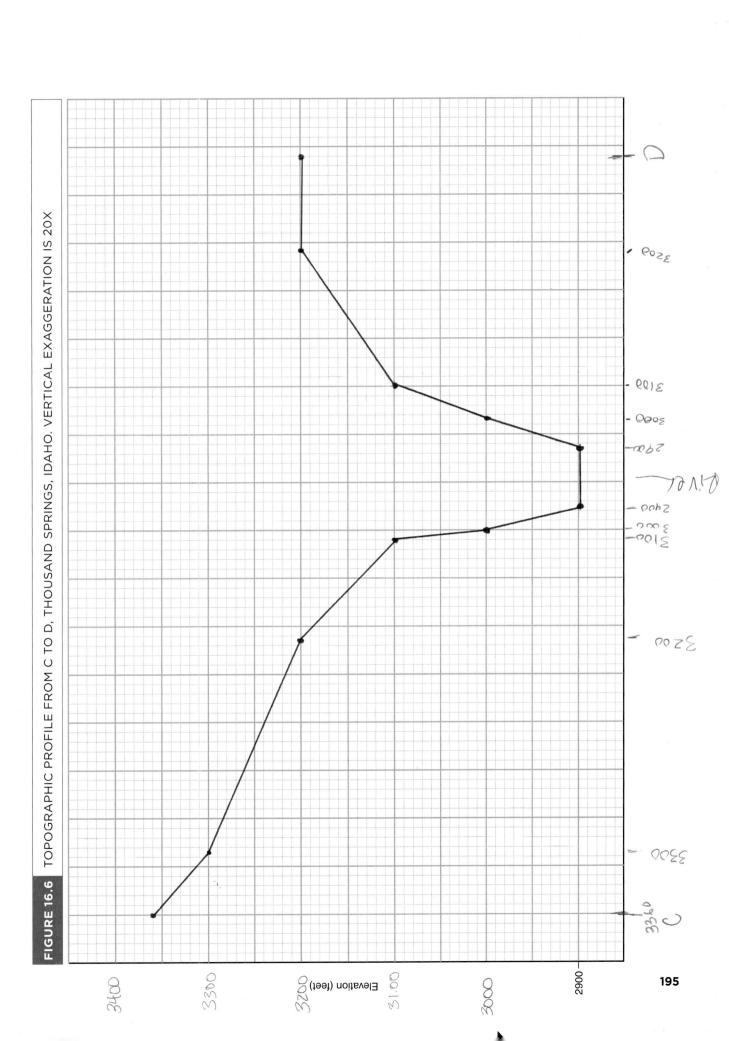

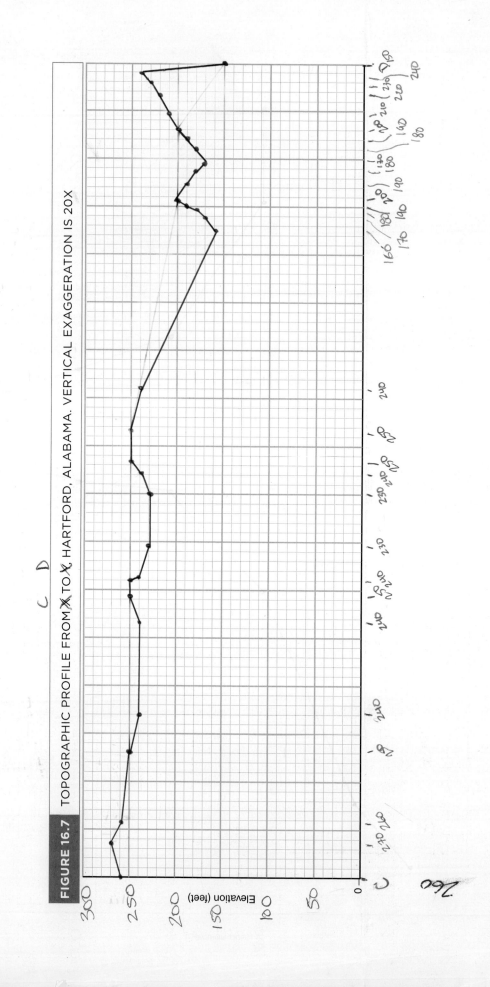

FIGURE 16.7 TOPOGRAPHIC PROFILE FROM X TO X, HARTFORD, ALABAMA. VERTICAL EXAGGERATION IS 20X

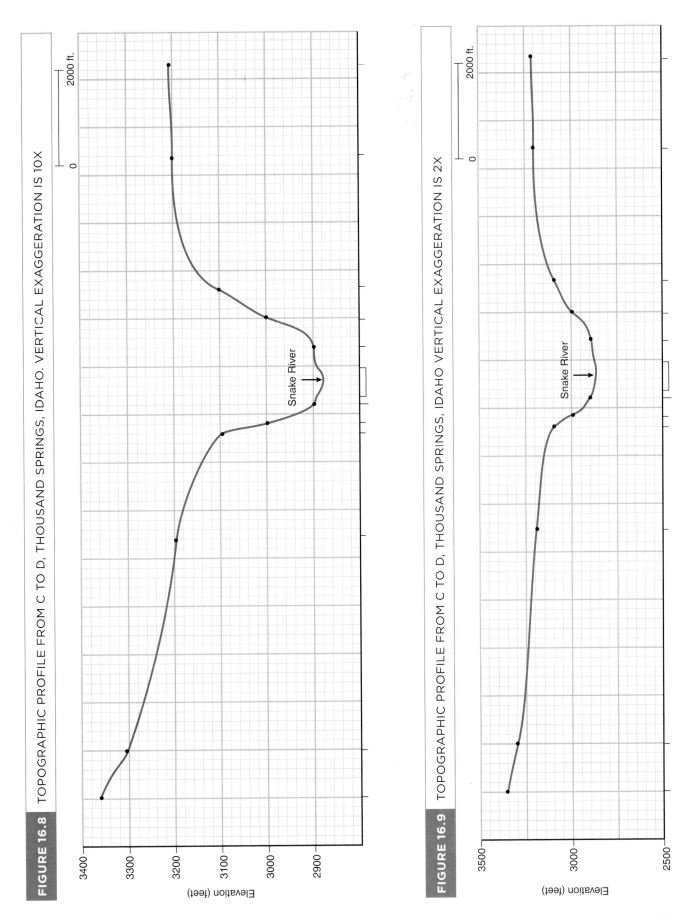

FIGURE 16.8 TOPOGRAPHIC PROFILE FROM C TO D, THOUSAND SPRINGS, IDAHO. VERTICAL EXAGGERATION IS 10X

FIGURE 16.9 TOPOGRAPHIC PROFILE FROM C TO D, THOUSAND SPRINGS, IDAHO. VERTICAL EXAGGERATION IS 2X

10. Use the topographic profiles you drew (Figures 16.6 and 16.7) to determine the local relief for the following two river valleys:

	Snake River, Idaho	Hurricane Creek, Alabama
a. Highest elevation:	3660 ft	270 ft
b. Lowest elevation:	2900 ft	160 ft
c. Local relief:	760 ft	110 ft

11. Examine the contour line that crosses Hurricane Creek on the Hartford, Alabama, map (Appendix E, Figure E.8) between points E and F, marked with stars (*).

    a. Which way does this contour line bend (north, south, northeast, etc.) as it crosses Hurricane Creek?

    To the south

    b. Based on your answer to part (a), toward what direction (north, south, northeast, etc.) is the water in Hurricane Creek flowing?

    North to South

    c. Examine the tributaries on the western side of Hurricane Creek. Do the contour lines crossing these tributaries to Hurricane Creek bend toward or away from Hurricane Creek?

    bend towards Hurricane Creek

    d. Based on your answer to part (c), is the water in these tributaries flowing into or out of Hurricane Creek?

    Flowing Into and out of Hurricane Creek?

    e. Keeping in mind your answers to the above questions, which of the two lines labeled G and H on the Hartford, Alabama, map is a more likely location for a tributary to Hurricane Creek? Why?

    line H, because it's close to 4 differents tributory currents.

12. Daughtry Pond, located south of the city of Hartford (Appendix E, Figure E.8), is colored blue, which lets you know it's a body of water. If the pond weren't there, how could you tell this is a low spot on the landscape?

    because the contour lines are small, showing lower elevation.

13. If you drew a profile along the line from K to L on the Hartford map (Appendix E, Figure E.8), would the land between points K and L be lower or higher than 250 feet? How do you know?

    The land will be lower than 250 ft. because if I look closely I can see the next contour line go down 10 feet from the letter K to the letter L.

14. The profile that you drew in Figure 16.7 ended at point D on the Hartford map (Appendix E, Figure E.8). The 250-foot contour line that runs through point D is a closed contour line. What would be a good estimate of the elevation of the land inside of this 250-foot contour line? Why?

15. The small squares on the Hartford map (Appendix E, Figure E.8) represent individual buildings (houses, schools, businesses, etc.).

    a. In what part of Hartford are there the most new buildings?

    b. Why aren't all the buildings in downtown Hartford drawn in?

    c. How does the map show that the downtown area has a lot of buildings?

16. Which map has more wooded area, Hartford, Alabama, or Thousand Springs, Idaho? How do you know?

17. Streams and rivers are shown in blue. Streams that flow only during the wet season or during storms are called intermittent streams and are shown by a blue line drawn as a dash and three dots. Streams that flow all year round are shown by a solid blue line. Larger streams and rivers are thicker blue lines.

    a. Which river probably has more water in it, the Snake River in Idaho, or Hurricane Creek in Alabama? How do you know?

    b. Do all the tributaries to Hurricane Creek flow all year round? How do you know?

Name: _____   Section: _____

PART 3 — MAP SCALE

Use the Chief Mountain, Montana-Alberta (Appendix E, Figure E.9) and the Saint Mary, Montana-Alberta (Appendix E, Figure E.10) maps to answer the following questions. The two maps show the same region of Montana.

1. What is the scale for each of these maps?
   Chief Mountain   __1:24,000__
   Saint Mary   __1:100,000__

2. What is the contour interval for each of these maps?
   Chief Mountain   __40__ feet
   Saint Mary   __50__ meters = _____ feet

3. What is the total relief (difference between highest and lowest elevations) for each of these maps?

	Chief Mountain	**Saint Mary**
a. Highest elevation on the map:	9080 ft	2858 m.
b. Lowest elevation on the map:	4720 ft	1100 m.
c. Total relief:	= 4,360 ft	= 1,758 m × 3.28 ft/meters = 5,766.24 feet

4. Why is the contour interval on the Saint Mary map so much larger than on the Chief Mountain map (compare the intervals in feet)? (Hint: Consider the differences in map scale.) Because of the Saint Mary deep terrain. and a lot of topographic relief.

5. Comparing distances:

   a. Use a ruler to measure how many inches separate point A, the benchmark on Chief Mountain, to the benchmark at point B, on the two maps. It doesn't matter that the Saint Mary map has metric units; you can still measure the distance in inches. This distance is the *map distance* separating these two points.

   Chief Mountain map distance   __8"__ inches
   Saint Mary map distance   __2"__ inches

   b. On which map is the map distance longer? __Chief Mountain__

   c. Is the map with the longer map distance the smaller or the larger scale map? __the smaller scale__

   d. Convert these map distances to actual distances on the earth's surface in miles using the representative fraction. Even though the Saint Mary map has elevations in metric units, any type of measurement may be used with a representative fraction. Show all your work below.   12 in = 1 ft   5280 ft = 1 mi

   Chief Mountain: 8" × 24,000 = 192,000 ÷ 12/1ft = 16,000 ft ÷ 5280 = 3.03 miles

   Saint Mary: 2" × 100,000 = 200,000 ÷ 12/1ft = 16,666 ft ÷ 5280 = 3.15 miles
   ft to 1 mil

1 mile = 5280

Chief Mountain earth distance   __3.03__ miles

Saint Mary earth distance   __3.15__ miles

e. Did you get the same (or at least close) answers for the actual distance on the earth's surface for part (d)? If your answers are not close, you probably made a mistake in measurement or calculation. Find your mistake and correct it.

6. Figures 16.10 and 16.11 show topographic profiles from point A, the benchmark on Chief Mountain, to the benchmark at point B on the two maps; that is, the two profiles show the same landscape although they are drawn from different maps. Both profiles have a 2X vertical exaggeration.

   a. Although the two profiles show the same landscape and have the same vertical exaggeration, one profile is much larger (both longer and taller) than the other. Why?

   Because F

   b. If Figure 16.11, the profile from the Saint Mary map, had been drawn with a 10X vertical exaggeration, it would be approximately as tall (take up as much vertical space on the Y-axis) as the profile from the Chief Mountain map shown in Figure 16.10. If we want to compare these two profiles, would it be better to compare the two profiles shown, or to redraw the Saint Mary profile so it's about as tall as the Chief Mountain profile? Why?

   c. Which profile provides more information on exactly what the topography would look like to a hiker traveling along this profile line?

7. a. Under what circumstances, or for what purposes, would the Chief Mountain map (Appendix E, Figure E.9) and its profile (Figure 16.10) be the most useful of the two maps?

   b. Under what circumstances, or for what purposes, would the Saint Mary map (Appendix E, Figure E.10) and its profile (Figure 16.11) be the most useful of the two maps?

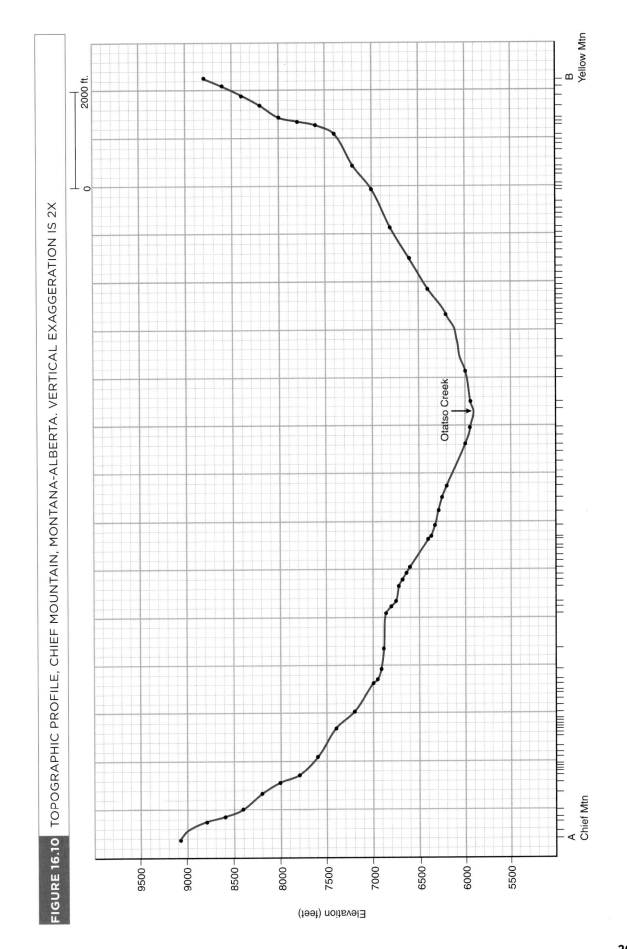

FIGURE 16.10 TOPOGRAPHIC PROFILE, CHIEF MOUNTAIN, MONTANA-ALBERTA. VERTICAL EXAGGERATION IS 2X

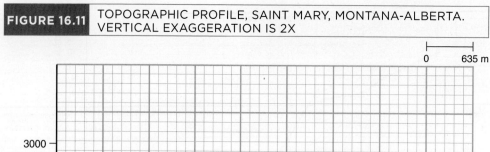

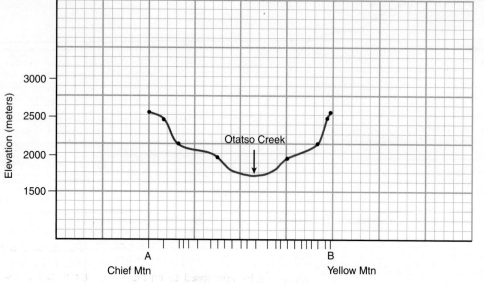

204

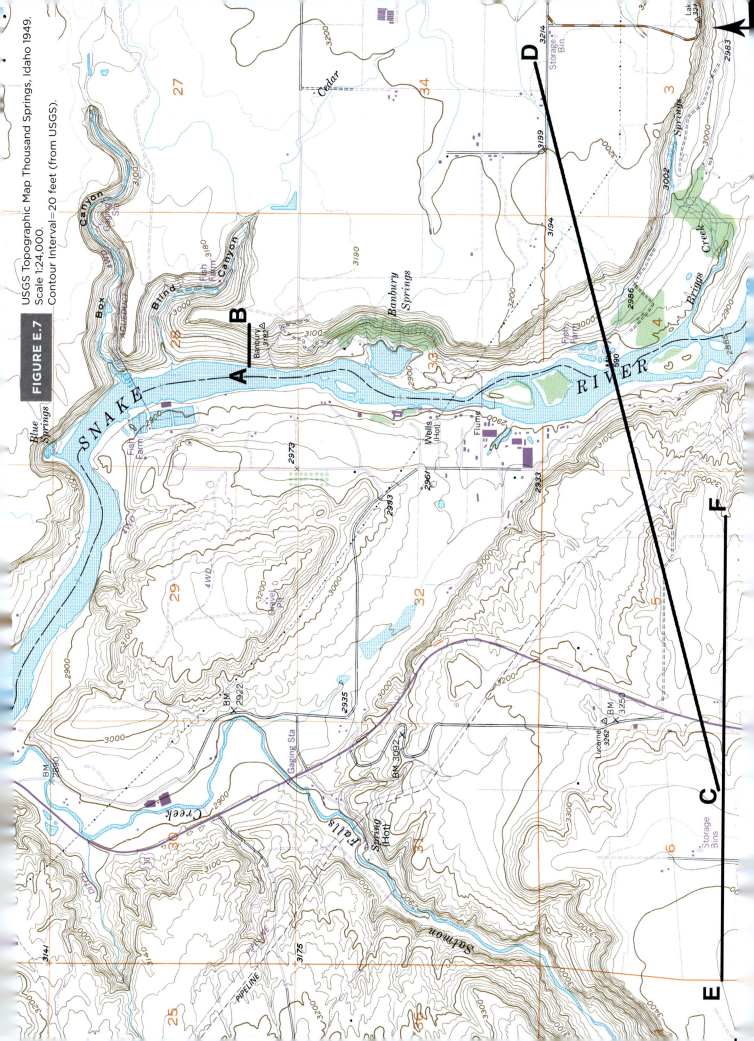

FIGURE E.7
USGS Topographic Map Thousand Springs, Idaho 1949.
Scale 1:24,000.
Contour Interval=20 feet (from USGS).

FIGURE E.6
Aerial Photograph of Devastation Trail Rainforest, 21 November 2001. Local time: 10:55. Scale=1:2,325.

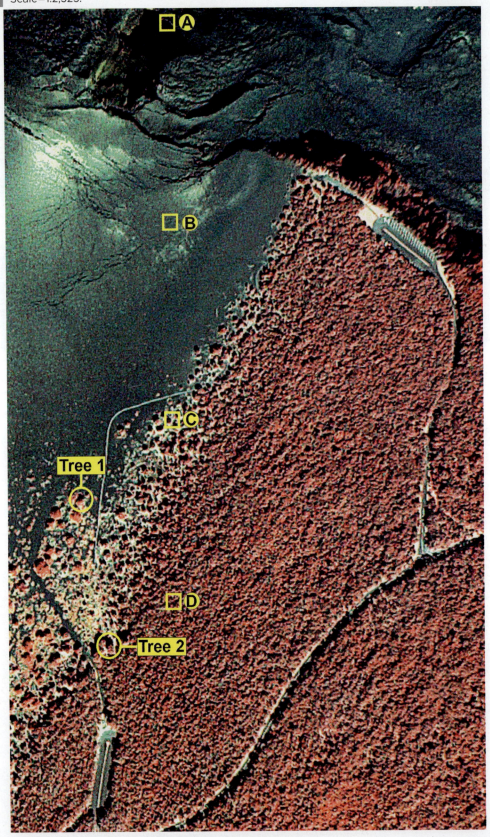

(NASA Ames Research Center: Orthophoto Kilauea Crater SE)

FIGURE E.8 USGS Topographic Map Hartford, Alabama 1980. Scale 1:24,000. Contour Interval=10 feet (from USGS).

FIGURE E.9 USGS Topographic Map Chief Mountain, Montana-Alberta 1968. Scale 1:24,000. Contour interval=10 feet (from USGS).